# *Writing Effective Work Proposals*
## 有效撰寫英文工作提案

By Ted Knoy
柯泰德

Illustrated by Wang Min-Chia
插圖：王敏嘉

*Ted Knoy is also the author of the following books in the Chinese Technical Writers' Series*（科技英文寫作系列）*and the Chinese Professional Writers' Series*（應用英文寫作系列）:

An English Style Approach for Chinese Technical Writers
精通科技論文（報告）寫作之捷徑

English Oral Presentations for Chinese Technical Writers
作好英語會議簡報

A Correspondence Manual for Chinese Technical Writers
英文信函參考手冊

An Editing Workbook for Chinese Technical Writers
科技英文編修訓練手冊

Advanced Copyediting Practice for Chinese Technical Writers
科技英文編修訓練手冊──進階篇

Writing Effective Study Plans
有效撰寫英文讀書計畫

This book is dedicated to my wife, Hwang Li Wen.

# 序　言

　　國人寫英文工作提案的機會愈來愈多，尤其在台灣加入世界貿易組織（WTO）之後，英文寫作不僅是溝通的工具，更是競爭力的表現。可惜國人往往無法有效利用所學英文來主導工作，無法用英文明確界定工作提案的最終目的，結果寫出的英文工作提案結構不全，甚至在計畫結束時工作提案還未完成。如此，多年所學英文便無法有效利用在工作上，實在可惜。

　　本書重點放在如何在工作計畫一開始時便可以用英文來規劃整個工作提案，由工作提案的背景、行動、方法及預期的結果漸次教導國人如何寫出具有良好結構的英文工作提案。如此用英文明確界定工作提案的程序及工作目標更可以確保英文工作提案及工作計畫的即時完成。對工作效率而言也有助益。

　　在國人積極加入WTO之後的調整期，優良的英文工作提案寫作能力絕對是一項競爭力快速加分的工具。

唐傳義　主任<br>
國立清華大學<br>
資訊工程學系

# Table of Contents

# Foreword

Professional writing is essential to the international recognition of Taiwan's commercial and technological achievements. "The Chinese Professional Writers' Series" seeks to provide a sound English writing curriculum and, on a more practical level, to provide Chinese professionals with valuable reference guides. The Series supports Chinese professional writers in the following areas:

## Writing style

The books seek to transform old ways of writing into a more active and direct writing style that better conveys an author's main ideas.

## Structure and content

The series addresses the organization of the structure and content of reports and other common forms of writing.

## Quality

Inevitably, writers prepare reports to meet the expectations of editors, referees/reviewers, as well as to satisfy the requirements of journals. The books in this Series are prepared with these specific needs in mind.

Writing Effective Work Proposals is the second book in the Chinese Professional Writers' Series.

*Writing Effective Work Proposals*為「應用英文寫作系列（The Chinese Professional Writers' Series）」之第二本書，本書中練習題部分主要是幫助國人進階學習的申請及避免，糾正常犯寫作格式上錯誤，由反覆練習中，進而熟能生巧提昇寫作及編修能力。

　　「應用英文寫作系列」將針對以下內容逐步協助國人解決在英文寫作上所遭遇之各項問題：

A.寫作型式：把往昔通常習於抄襲的寫作方法轉換成更積極主動的寫作方式，俾使讀者所欲表達的主題意念更加清楚。更進一步糾正國人寫作口語習慣。

B.方法型式：指出國內寫作者從事英文寫作或英文翻譯時常遇到的文法問題。

C.內容結構：將寫作的內容以下面的方式結構化：目標、一般動機、個人動機。並瞭解不同的目的和動機可以影響報告的結構，由此，獲得最適當的報告內容。

D.內容品質：以編輯、審查委員的要求來寫作此一系列之書籍，以滿足讀者的英文要求。

# Introduction

This writing workbook aims to instruct students in writing a well-structured work proposal. The following elements of an effective work proposal are introduced: outlining the first part of the work proposal by describing the project background; outlining the second part of the work proposal by describing the approach to solving the problem; writing the problem statement; writing the hypothesis statement; writing the abstract which summarizes the background, objective(s), methodology, anticipated results and the overall contribution of the work to a particular field and, finally, combining everything into a concise work proposal.

Many authors admit that they do not begin writing until the rest of the project is completed because they view writing as an administrative task that involves reflecting upon something having occurred in the past, rather than upon something which is unfolding at the time of writing. The thought that the actual writing could direct the project or research process is as alien as the notion that the report or research article could be written while the rest of the project itself approaches completion. Such authors fail to realize that identifying the background and goals of a project beforehand ensures not only that writing is completed when the project or research is completed, but also that the report is well structured. Carefully identifying the background and goals of the project beforehand also ensures that the readers' interests are properly addressed. Each unit in this workbook represents a series of steps that an author can make not only to write a concise work proposal, but also to lay out a clear structure for the final report or article.

Each unit begins with three visually represented situations that provide essential information to help students to write a specific part of a work proposal. Additional written activities, relating to those three situations, help students to understand how the visual representation relates to the ultimate goal of writing an effective work proposal. An Answer Key makes this book ideal for classroom use. For instance, to test a student's listening comprehension, a teacher can first read the text that describes the situations for a particular unit. Either individually or in small groups, students can work through the exercises to produce a well-structured work proposal.

# 簡　介

　　本書主要教導讀者如何建構良好的英文工作提案。書中內容包括：1. 工作提案計畫（第一部分）：背景 2. 工作提案計畫（第二部分）：行動 3.問題描述 4.假設描述 5. 摘要撰寫（第一部分）：簡介背景、目標及方法 6. 摘要撰寫（第二部分）：歸納希望的結果及其對特定領域的貢獻 7.綜合上述寫成精確工作提案。

　　許多國人都是在工作方案完成時才開始撰寫相關英文提案，他們視撰寫提案為行政工作的一環，只是消極記錄已完成的事項，而不是積極的規劃掌控未來及現在正進行的工作。如果國人可以在撰寫英文提案時，事先仔細明辨工作計畫提案的背景及目標，不僅可以確保寫作進度、寫作結構的完整性，更可兼顧提案相關讀者的興趣強調。本書中詳細的步驟可指導工作提案寫作者達成此一目標。

　　書中的每個單元呈現三個視覺化的情境，提供國人英文工作提案寫作實質訊息，而相關附加的寫作練習讓讀者做實際的訊息運用。此外，本書也非常適合在課堂上使用，教師可以先描述單元情境而讓學生藉由書中練習循序完成具有良好架構的工作提案。

# *Unit One*

 Larry    Information Management( 資訊管理 )

Intranets

enterprises

network management systems

hardware
software

 Julie    Distance Learning( 遠距教學 )

Teacher    Students

Effectiveness ??

## Outlining the work proposal (part one): describing the project background

### 工作提案計畫（第一部分）：背景

**Vocabulary and related expressions**　相關字詞

Intranets　企業內網路
conventional network management systems　慣用的網路管理系統
implement　工具
upgrading hardware and software　軟硬體的升級
distance learning courses　遠距學習課程
lack　缺少
feasible strategies　適當的方法
reliable assessment methods　可靠的評價方式
inhibits　抑制
mathematics courses　數學課程
conventional editors　傳統編輯器
plain text　純文字
user interface　使用者介面
complex mathematical equations　複雜的數學方程式
mathematical symbols　數學符號

**A** Write down the key points of the situations on the preceding page while the instructor reads aloud the script on page 320.

**Situation 1**

_____

_____

_____

**Situation 2**

_____

_____

_____

**Situation 3**

_____

_____

_____

**B** Based on the three situations in this unit, write three questions beginning with **What**, and answer them.

**Examples**

*What inhibits distance learning?*

*The lack of reliable assessment methods*

*What do many enterprises use to accelerate commercial activities?*

*Intranets*

1. _____

   _____

2. _____

   _____

3. _____

   _____

## C Based on the three situations in this unit, write three questions beginning with *Which*, and answer them.

**Examples**

*Which problem is Julie trying to solve?*

*Many distance learning courses lack feasible strategies to assess student performance.*

*Which general topic is Larry concerned with?*

*Many enterprises use Intranets to accelerate commercial activities.*

1. _____

_____

2. _____

_____

3. _____

_____

**D** Based on the three situations in this unit, write three questions beginning with **Why**, and answer them.

**Examples**

*Why is distance learning inhibited?*

*Because of the lack of reliable assessment methods*

*Why does Larry think that conventional network management systems can not be implemented in a typical enterprise's Intranet?*

*Because they are too expensive and complicated*

1. _____

   _____

2. _____

   _____

3. _____

   _____

## E  Write questions that match the answers provided.

**Examples**

*What are many enterprises using to accelerate commercial activities?*

*Intranets*

*What amount is upgrading a network management system's hardware and software too expensive?*

*300 thousand U.S. dollars*

1. _____

_____

Because they are too expensive and complicated

2. _____

_____

On the Web

3. _____

_____

Conventional ones, using plain text as the user interface

# Unit One

Outlining the work proposal (part one):
describing the project background

## 工作提案計畫（第一部分）：背景

1 Setting of work proposal 工作提案建構
2 Work problem 工作問題
3 Quantitative specification of problem 問題的量化
4 Importance of problem 問題的中心

Outlining the work proposal begins with describing the project background, which consists of the following parts.

**1. Setting of work proposal 工作提案建構. What is the topic of interest with which your work proposal is concerned? Can you point the reader of your work proposal to the general area with which you are concerned? 你工作提案的主題是什麼? 你的讀者可以明瞭工作提案的內容嗎?**

Consider the following examples:

◎Problems (bugs) detected by testing device drivers are commonly submitted to a database to analyze and compile statistics on all bugs.

◎The quality of wafers is essential, especially in light of the shrinking of electronic devices and the complexity of circuits in a chipset. The test period is prolonged to ensure that most functional paths are detected before mass production is undertaken.

◎Stereo 3D displays play an increasingly prominent role in computer graphics, visualization and virtual-reality systems.

◎Engineers often decide whether to adjust a manufacturing process according to measurement data. The quality of the measurement data is related to the measurement system and its environment.

◎Despite a large frequency offset, wide range locking with fast acquisition circuit-designed carrier recovery helps a digital receiver to lock the carrier frequency in a relatively short time and with a tolerable error rate.

◎ The decision-making environment is increasingly complex.

◎Consumer demand on the water supply in southern Taiwan has significantly increased in recent years owing to the growth of population and industry as well as elevated living standards.

◎Enterprises trade with each other by carefully considering business ratings to reduce investment risks.

◎Statistical approaches such as scored methods are extensively used to optimize the process parameters of ordered categorical responses.

◎Process quality and delivery time have been increasingly emphasized by industry.

◎More Internet-based learning environments are available to enhance knowledge construction.

◎More wafer fabs use a control chart to detect assignable causes, making the effective control of the wafer process by engineers extremely difficult.

◎Predicting the amount of residuals of ground water contaminants has been extensively studied.

**2. Work problem 工作問題. What is the problem that you are trying either to solve or to more thoroughly understand in the work proposal? 你的工作提案裡有你試著要解決或是想更進一步瞭解的問題嗎?**

Consider the following examples.

◎Conventional methods fail to establish the stability of a measurement system, specify how to repeat, reproduce and match a measurement system, or provide a criterion of judgment.

◎The inability to monitor online whether products are out of specification increases the likelihood of poor wafer quality.

◎In a high speed PC, the resistor impedance, crosstalk, power quality, and EMI constitute bottlenecks for further development.

◎Conventional analysis models can not accurately forecast capacity in the short (dynamic) and long (static) term owing to the varying complexity of the model(s) employed.

◎Conventional recognition approaches can not ensure individual security when commercial transactions are conducted on the Internet.

◎ Conventional methods cannot solve multi-state problems.

◎Students spend much time in coding programs for exercises when learning GAs, making it impossible to implement many GAs in a relatively short time.

◎The critical behavior of materials, described by the weakly diluted O(m) model, is unclear with respect to the dimensions of the order parameter.

◎Fully understanding the relationship between microscopic description and macroscopic phase behavior in a binary liquid system is extremely difficult.

◎The exact process by which radiation impacts ionic conductivity remains unknown.

◎The physical nature of the dielectric properties of DMAGaS-DMAAlS ferroelectrics remains unclear.

◎The theoretical description of exoemission from alkali halide crystals is insufficient, making the analysis and interpretation of exoemission current-related data difficult.

◎Exactly why the RSOE mechanism is present in different materials remains unknown.

◎Whether radiation decay of I-V dipoles occurs in CsJ crystals remains unknown.

◎The IMD Competitiveness Simulations neglect interactions among related factors.

**3. Quantitative specification of problem 問題的量化. How can you quantitatively define the problem so that the reader understands the quantitative limitations of previous efforts to solve it? 你要如何量化問題來讓你的讀者明白之前文獻研究所遇到的量化限制？**

Consider the following examples.

◎Although increasing the wafer size and reducing the device' s dimensions are strongly desired, a deep submicron process, 0.15um or lower will decrease not only the yield and quality of the product, but also the company's technical and market position.

◎The inaccuracy of wireload models' determination of wire delays ranges from roughly 10% to 30%.

◎The CMOS ROM capability in BIOS code is only 256k. Therefore, more codes must be added to increase the number of functions.

◎A loop filter with a wide bandwidth experiences a large vibration and ultimately a high error rate, whereas a loop filter with a narrow bandwidth leads to slow convergence that takes more than ten times longer than the estimated acquisition time.

◎The two view-points stereo display requires more than twice as much geometrical computation time than does the non-stereo 3D display.

◎Piecewise linearization algorithms that require more than 10 hours to obtain the optimal solution to a general nonlinear programming problem make equipment-related costs too high.

◎A prediction accuracy of only 60% among previous models exposes traders to unnecessarily high risks.

◎Conventional models require seven steps to construct a 3D model and have error rates that exceed 5%, making it impossible to widely commercialize animation software and construct realistic digital objects.

◎An error rate that exceeds 20% does not meet the R.O.C.'s environmental protection standards.

◎A situation in which the formability near room temperature does not reach 20% elongation limits the commercial applications of magnesium and its alloys.

◎If the geometric parameters can not reach an accuracy of 5%, engineers can neither predict the fatigue life nor enhance the yield of a flip chip package.

◎For a solution that varies with problem states in a multi-state problem,···

◎Students who can implement only one GA in two weeks will learn GAs less effectively than those who can implement more .

◎The lack of uniform data on various dimensions of an order parameter makes it impossible to obtain the number of marginal dimensions that distinguish the above scenarios.

◎The failure to understand precisely which microscopic features form a particular phase topology makes it impossible to analyze from first principles the behavior of simple model thermodynamic functions near the critical point .

◎Increasing the evaporation speed by 1% increases the resistance by 5%.

**4. Importance of problem 問題的中心.  How will leaving the problem unsolved or not thoroughly understood negatively impact the reader of your work proposal?  如果問題沒被解決或是充分瞭解，這對工作提案的讀者會有多大的負面衝擊?**

Consider the following examples.

◎Such a high cost prevents us from verifying the designs of platforms, and the low speed makes testing difficult and the results unrealistic.

◎Pricing interest rate derivatives by adopting these inappropriate fitting forward rate methods will lead to a large deviation from market data.

◎Neglecting the assigned locations of the cells during synthesis makes it impossible to evaluate the wire delays and achieve optimization efficiently.

◎A reduction in power quality causes the DDR_SDRAM to perform unsatisfactorily. Moreover, resistor impedance and cross-talk cause noise in the PCB, negatively impacting the VGA Card's speed during operation and falling below FCC standards with respect to EMI.

◎The inability to reduce the visible, white smoke and odor to acceptable levels not only creates a potentially unsafe working environment, but also poses a perceived threat to the nearby community.

◎Equipment-related costs are too high.

◎The equal participation of students can not be ensured.

◎The inability of PDA users to receive updated information in a timely manner will limit the use of PDAs to within a narrow range.

◎Engineers can neither accurately predict the fatigue life nor enhance the yield of a flip chip package.

◎Conventional methods neglect the multi-state property and thus yield an inaccurate and unfeasible solution.

◎The effectiveness of learning GAs will be diminished.

◎Conventional models may cause decision makers to select an inappropriate scheme.

◎Their implementation leads to an inaccurate optimal combination of process parameters.

◎The lack of an objective procedure leads to inefficiency and a high overhead cost.

◎Using such charts often leads to ineffective and futile searches for assignable causes, wasting considerable amounts of manpower and capital.

**F** In the space below, outline the first part of your work proposal by describing the project background.

Setting of work proposal 工作提案建構:

_____

_____

_____

Work problem 工作問題:

_____

_____

_____

Quantitative specification of problem 問題的量化:

_____

_____

_____

Importance of problem 問題的中心:

_____

_____

_____

## G Look at the following examples of outlining the first part of the work proposal by describing the project background.

*Statistics* 統計相關分類

**Setting of work proposal** 工作提案建構 Decision-making environments are increasingly complex. **Work problem** 工作問題 However, conventional evaluation models are concerned only with economic factors and neglect factors that can not be evaluated in money. **Importance of problem** 問題的中心 Therefore, conventional models may cause decision makers to select an inappropriate scheme.

*Civil Engineering* 土木相關分類

**Setting of work proposal** 工作提案建構 Consumer demand on the water supply in southern Taiwan has significantly increased in recent years owing to the growth of population and industry, as well as elevated living standards. **Work problem** 工作問題 Although the reservoir supply in southern Taiwan can satisfy the current demand for water, increasing demand will soon surpass the reservoir's capacity. **Importance of problem** 問題的中心 Therefore, the lack of an effective distribution strategy for water from current reservoirs will lead to frequent water shortages in southern Taiwan.

*Statistics* 統計相關分類

**Setting of work proposal** 工作提案建構 Statistical approaches such as scored methods are extensively used to optimize the process parameters of ordered categorical responses. **Work problem** 工作問題 However, these methods inaccurately estimate the dispersion effect, confounding the location and dispersion effects. **Importance of problem** 問題的中心 Therefore, their implementation leads to an inaccurate optimal combination of process parameters.

*Industrial Engineering* 工業工程相關分類

**Setting of work proposal** 工作提案建構 Process quality and delivery time have been increasingly emphasized by industry. **Work problem** 工作問題 However, conventional process capability indices (PCIs) can neither objectively assess quality and delivery time nor identify the relationship between PCIs and yield rate. **Importance of problem** 問題的中心 Therefore, the lack of an objective procedure leads to inefficiency and a high overhead cost.

*Industrial Engineering* 工業工程相關分類

**Setting of work proposal** 工作提案建構 A process capability index is a highly effective means of evaluating process performance. **Work problem** 工作問題 However, conventional process capability indices can not accurately measure a non-normal distribution process. **Quantitative specification of problem** 問題的量化 The inability to measure accurately a non-normal distribution process **Importance of problem** 問題的中心 leads to error when comparing processes or selecting an alternative supplier.

*Statistics* 統計相關分類

**Setting of work proposal** 工作提案建構 Enterprises trade with each other by carefully considering business ratings to reduce investment risks. **Work problem** 工作問題 However, conventional mathematical models have difficulty in discriminating between multiple ranks. **Quantitative specification of problem** 問題的量化 A prediction accuracy of only 60% among past models **Importance of problem** 問題的中心 exposes traders to unnecessarily high risks.

*Civil Engineering* 土木相關分類

**Setting of work proposal** 工作提案建構 Demands on the water supply in Taiwan have skyrocketed owing to the island's increasing population and elevated living standards. **Work problem** 工作問題 Eventually, the current reservoir supply will be insufficient if current personal and industrial demand persists. **Importance of problem** 問題的中心 Consequently, this surging demand will lead to the  increased drilling of new wells to locate groundwater resources, which is extremely expensive.

*Industrial Engineering* 工業工程相關分類

**Setting of work proposal** 工作提案建構 More wafer fabs use a control chart to detect assignable causes, making the effective control of the wafer process by engineers extremely difficult. **Work problem** 工作問題 However, conventional control charts are designed for manufacturing systems with only one source of variation, making it impossible to control several independent sources of variation. **Importance of problem** 問題的 中心 Using such charts often leads to ineffective and futile searches for assignable causes, expending considerable amounts of manpower and capital.

*Civil Engineering* 土木相關分類

**Setting of work proposal** 工作提案建構 Predicting the amount of residuals in ground water contaminants has been extensively studied. **Work problem** 工作問題 However, uncertain and insufficient hydrology data make it impossible to accurately predict their amount. **Quantitative specification of problem** 問題的量化 An error rate that exceeds 20% **Importance of problem** 問題的中心 does not meet the R.O.C.'s environmental protection standards.

*Civil Engineering* 土木相關分類

**Setting of work proposal** 工作提案建構 Groundwater usage in Taiwan is increasing at an accelerated rate, **Work problem** 工作問題 leading to a growing incidence of groundwater pollution in major groundwater supply regions in Taiwan, such as the Ping Tung Plain. **Importance of problem** 問題的中心 Contaminants that pollute the aquifer make drinking the water from that groundwater source impossible for several years.

*Distance Learning* 遠距教學相關分類

**Setting of work proposal** 工作提案建構 More Internet-based learning environments are available to enhance knowledge construction. **Work problem** 工作問題 However, learning activities of these learning environments over rely on in-group cooperation, erroneously implying that group members participate equally. **Importance of problem** 問題的中心 Therefore, the inability of cooperation and competition to complement each other in an Internet-based learning environment makes it impossible to ensure that students participate equally.

*Teacher Education* 教育相關分類

**Setting of work proposal** 工作提案建構 Taiwan's Ministry of Education is increasingly emphasizing the use of multi-assessment in middle schools and universities. **Work problem** 工作問題 However, conventional methods of assessing students' abilities fail to assess higher-order thinking owing to their inability to motivate students properly. **Importance of problem** 問題的中心 Therefore, Taiwan's educational system cannot multi-assess the higher-order thinking of students.

*Distance Learning* 遠距教學相關分類

**Setting of work proposal** 工作提案建構 The benefits of web learning are widely recognized by educators and researchers. **Work problem** 工作問題 However, recent studies have suggested that many web-learning activities merely support the memorization of facts, passive learning, or even disenchanted browsing. **Importance of problem** 問題的中心 Therefore, the inability of web collaborative design activity to promote students' higher order thinking makes it impractical to implement such an innovative means of instruction.

*Computer Science* 資訊科學相關分類

**Setting of work proposal** 工作提案建構 Piecewise linearization algorithms are extensively used in nonlinear programming. **Work problem** 工作問題 However, these algorithms require too much time to obtain an optimum solution. **Quantitative specification of problem** 問題的量化 For instance, piecewise linearization algorithms that require more than ten hours to obtain the optimal solution for general nonlinear programming problems **Importance of problem** 問題的中心 make equipment-related costs too high.

*Information Management* 資訊管理相關分類

**Setting of work proposal** 工作提案建構 3D models are extensively adopted in multimedia applications owing to their relatively low cost and ease with which they can construct animated 3D objects. **Work problem** 工作問題 However, conventional 3D models are too time-consuming and inaccurate when constructing digital objects since they manually retrieve 2D images. **Quantitative specification of problem** 問題的量化 For instance, a situation in which conventional models require seven steps to construct a 3D model with an error rate of over 5% **Importance of problem** 問題的中心 makes it impossible to widely commercialize animation software and construct realistic digital objects.

*Information Management* 資訊管理相關分類

**Setting of work proposal** 工作提案建構 Geographic-based information is increasingly used in daily living. **Work problem** 工作問題 However, the passive mode of accessing information fails to effectively transmit geographic-based information to PDA users. **Importance of problem** 問題的中心 Therefore, the inability of PDA users to receive updated information in a timely manner will limit the use of PDAs to within a narrow range.

23

*Computer Science* 資訊科學相關分類

**Setting of work proposal** 工作提案建構 Intelligent information retrieval systems utilize knowledge bases to increase the effectiveness of retrieval. **Work problem** 工作問題 However, most knowledge bases are constructed by inquiring of domain experts to acquire knowledge. **Importance of problem** 問題的中心 This inefficient approach requires, on average, one hour to construct a knowledge base of 50 concepts.

*Mechanical Engineering* 機械相關分類

**Setting of work proposal** 工作提案建構 Geometric parameters affect the reliability of flip chip packaging under thermodynamic loading, **Work problem** 工作問題 making it impossible to accurately predict geometric parameters of a C4 type solder joint. **Quantitative specification of problem** 問題的量化 Therefore, if the geometric parameters can not reach an accuracy of 5%, **Importance of problem** 問題的中心 then engineers can neither predict the fatigue life nor enhance the yield of a flip chip package.

*Computer Science* 資訊科學相關分類

**Setting of work proposal** 工作提案建構 Many machine learning and optimization application-related problems are solved by GAs that include the multi-state property. **Work problem** 工作問題 However, conventional methods cannot solve multi-state problems. **Quantitative specification of problem** 問題的量化 For a solution that varies with the problem state in a multi-state problem, **Importance of problem** 問題的中心 conventional methods neglect the multi-state property and thus yield an inaccurate and unfeasible solution.

*Computer Science* 資訊科學相關分類

**Setting of work proposal** 工作提案建構 An increasing number of Genetic Algorithm (GA) courses are offered to solve optimization problems. **Work problem** 工作問題 However, students spend much time in coding programs for exercises when learning GAs, making it impossible to implement many GAs in a relatively short time. **Quantitative specification of problem** 問題的量化 For instance, students who can implement only one GA in two weeks **Importance of problem** 問題的中心 will learn GAs less effectively than those who can implement more.

*Materials Science* 材料相關分類

**Setting of work proposal** 工作提案建構 Magnesium alloys have great potential for diverse use in the automotive, railway, aerospace, computer, communications and consumer electronics industries since they have the lowest density of all metallic structural materials. **Work problem** 工作問題 However, these alloys have a low formability near room temperature and are more expensive than plastics, aluminum, steel and cast iron.

**Quantitative specification of problem** 問題的量化 A situation in which the formability near room temperature does not reach 20% elongation **Importance of problem** 問題的中心 limits the commercial application of magnesium and its alloys.

*Physics* 物理相關分類

**Setting of work proposal** 工作提案建構 Weak disorder leads to two scenarios for the critical behavior of the O(m) model. **Work problem** 工作問題 However, the critical behavior of materials described by the weakly diluted O(m) model is unclear with respect to the dimensions of the order parameter. **Quantitative specification of problem** 問題的量化 Additionally, the lack of uniform data on the various dimensions of an order parameter **Importance of problem** 問題的中心 makes it impossible to obtain the number of marginal dimensions that distinguish the above scenarios.

*Physics* 物理相關分類

**Setting of work proposal** 工作提案建構 The interaction between binary mixture components leads to diverse phase behavior with respect to relative molecular sizes and the strengths of their interactions. **Work problem** 工作問題 However, fully understanding the relationship between microscopic description and macroscopic phase behavior in a binary liquid system is extremely difficult. **Quantitative specification of problem** 問題的量化

Additionally, the failure to understand precisely which microscopic features form a particular phase topology **Importance of problem** 問題的中心 makes impossible an analysis from first principles of the behavior of simple model thermodynamic functions near the critical point .

*Physics* 物理相關分類

**Setting of work proposal** 工作提案建構 As well known, radiation induces defect rebuilding. **Work problem** 工作問題 However, the exact process by which radiation impacts ionic conductivity remains unknown. **Importance of problem** 問題的中心 The inability to understand its impact on ionic conductivity makes it impossible not only to estimate the radiation defect rebuilding in cesium halides doped with OH-impurities, but also to measure the ionic conductivity of irradiated crystals.

*Physics* 物理相關分類

**Setting of work proposal** 工作提案建構 The dielectric properties of DMAGaS-DMAAlS ferroelectrics produce a variation in temperature and pressure behavior. **Work problem** 工作問題 However, the physical nature of dielectric properties of DMAGaS-DMAAlS ferroelectrics remains unclear. **Importance of problem** 問題的中心 The inability to fully understand its physical nature further delays experimental investigation and possible industrial applications of these crystals.

*Physics* 物理相關分類

**Setting of work proposal** 工作提案建構 The surface control method based on exoelectron emission (EEE), a precise and nondestructive relaxational method, can detect and predict the early stages of the destruction of materials. Additionally, the exoemission from alkali halide crystals has been extensively studied, theoretically and experimentally. **Work problem** 工作問題 However, the theoretical description of exoemission from alkali halide crystals is insufficient, making it difficult to analyze and interpret exoemission current-related data. **Importance of problem** 問題的中心 Therefore, the inability to fully understand the exoemission mechanism makes it impossible to detect and investigate defects in the surface layouts of the crystals.

*Physics* 物理相關分類

**Setting of work proposal** 工作提案建構 Evaporation methods enhance the formation of scattering centers and the electrical conductivity of thin metal films. **Work problem** 工作問題 However, increasing the evaporation speed increases the resistance of metal films. **Quantitative specification of problem** 問題的量化 For instance, increasing the evaporaion speed by 1% increases the resistance by 5%, **Importance of problem** 問題的中心 making it impossible to prevent the resistance of thin films from fluctuating with a varying speed of evaporation.

*Physics* 物理相關分類

**Setting of work proposal** 工作提案建構 As well known, ionizing irradiation damages material.  **Work problem** 工作問題 However, the cause of the RSOE mechanism in different materials remains unknown. **Importance of problem** 問題的中心 Additionally, the inability to fully understand the RSOE mechanisms in II-VI semiconductors makes it impossible to accurately predict the irradiation conditions and thus increase structural efficiency.

*Physics* 物理相關分類

**Setting of work proposal** 工作提案建構 Dipole relaxation arises in doped CsJ crystals.  **Work problem** 工作問題 However, whether radiation decay of I-V dipoles occurs in CsJ crystals remains unknown. **Importance of problem** 問題的中心 Additionally, the inability to fully understand the formation of radiation-stimulated defects in doped CsJ crystals makes it impossible to use these crystals in technological elements.

*Information Management* 資訊管理相關分類

**Setting of work proposal** 工作提案建構 Taiwan's global competitiveness ranking in the IMD World Competitiveness Scoreboard is falling.  **Work problem** 工作問題 The IMD Competitiveness Simulations neglect interactions among related factors. **Importance of problem** 問題的中心

Use of the IMD Competitiveness Simulations possibly causes policy makers to make an inaccurate decision.

*Quality Control* 品保相關分類

**Setting of work proposal** 工作提案建構 Product life cycles in the information industry are shortening. Increasing market share depends on timely delivery of quality products. **Work problem** 工作問題 However, good product quality requires considerable time spent on testing during research and development. **Quantitative specification of problem** 問題的量化 If testing time accounts for 20% of the total research and development time, **Importance of problem** 問題的中心 then R&D engineers can not focus on writing code, reducing product quality and delaying production.

*Electrical Engineering* 電子相關分類

**Setting of work proposal** 工作提案建構 Problems (bugs) detected by testing device drivers are commonly submitted to a database in order to analyze and compile statistics on all bugs. **Work problem** 工作問題 However, the "dBase" format only supports simple and small records initially. **Quantitative specification of problem** 問題的量化 The increasing number of bugs added has led to weekly or twice weekly corruption of our database, **Importance of problem** 問題的中心 making it impossible to repair the damaged database if records are lost during the development of the drivers.

*Quality Control* 品保相關分類

**Setting of work proposal** 工作提案建構 The quality of wafers is essential, especially in light of the shrinking of electronic devices and the complexity of circuits in a chipset. The test period is prolonged to ensure that most functional paths are detected before mass production. **Work problem** 工作問題 However, the rising cost of tests has not improved overall testing. **Quantitative specification of problem** 問題的量化 For example, doubling the test time decreases product fault coverage by only 5%. **Importance of problem** 問題的中心 The lack of major test parameters and methods makes it impossible to enhance testing.

*Computer Graphics* 電腦製圖相關分類

**Setting of work proposal** 工作提案建構 Stereo 3D displays play an increasingly prominent role in computer graphics, visualization and virtual-reality systems. **Work problem** 工作問題 However, conventional methods inefficiently derive stereoscopic images owing to the complex geometrical computation required. **Quantitative specification of problem** 問題的量化 For example, the two view-points stereo display requires more than double the geometrical computation time than does the non-stereo 3D display. **Importance of problem** 問題的中心 Thus, the inability to solve the problem of inefficiency makes high-quality and real-time stereo 3D applications impossible.

*Quality Control* 品保相關分類

**Setting of work proposal** 工作提案建構 Engineers often decide whether to adjust a manufacturing process according to measurement data. The quality of measurement data is related to the measurement system and its environment. **Work problem** 工作問題 However, conventional methods fail to establish the stability of a measurement system, specify how to repeat, reproduce and match a measurement system, or provide a criterion of judgment. **Importance of problem** 問題的中心 An unstable or inaccurate measurement system will yield errors in process control and judgment, making it inappropriate in analyzing a manufacturing process.

*Quality Control* 品保相關分類

**Setting of work proposal** 工作提案建構 A reliable wafer chipset must not only have complementary functions, but also be efficient. In particular, the ability to forecast the life of a chipset is a priority concern in wafer products' quality and reliability. **Work problem** 工作問題 However, conventional burn-in boards do not have complementary functions and are expensive. **Importance of problem** 問題的中心 The inability to forecast accurately the life cycle of a wafer chipset lowers a product's reliability.

*Electrical Engineering* 電子相關分類

**Setting of work proposal** 工作提案建構 Despite a large frequency offset, wide range locking with fast acquisition circuit-designed carrier recovery helps a digital receiver to lock the carrier frequency in a relatively short time and with a tolerable error rate. **Work problem** 工作問題 However, conventional methods that involve a digital phase-locked loop (PLL) circuit cannot do so since the loop filter unit inside the circuit is typically a one-order low-pass filter. **Quantitative specification of problem** 問題的量化 Under such circumstances, a loop filter with a wide bandwidth experiences a large vibration and ultimately a high error rate., whereas a loop filter with a narrow one leads to slow convergence that takes more than ten times longer than the estimated acquisition time. **Importance of problem** 問題的中心 Furthermore, a loop filter with a narrow bandwidth may not recover the carrier while the receiver suffers from a large frequency or phase offset, leading to failure in the digital receiver.

*Electrical Engineering* 電子相關分類

**Setting of work proposal** 工作提案建構 The current trend of decreasing the length of a metal line in advanced IC processes has increased the importance of the gap fill ability of HDP-CVD in the IC backend process. **Work problem** 工作問題 However, the conventional HDP-CVD process is characterized by high Ar gas flow and pressure. **Quantitative specification of problem** 問題的量化 A situation in which Ar gas flow and pressure exceed acceptable levels **Importance of problem** 問題的中心 causes re-deposition in the high aspect ratio structure and decreases the gap fill ability.

*Quality Control* 品保相關分類

**Setting of work proposal** 工作提案建構 The product life of wafers is gradually becoming shorter as the conductor line width in integrated circuits decreases from the sub-micro to the deep sub-micro level. **Work problem** 工作問題 As the reliability and lifetime of wafer products continue to fall, **Importance of problem** 問題的中心 production falls below customers' specifications.

*Electrical Engineering* 電子相關分類

**Setting of work proposal** 工作提案建構 Consumer demand for stronger functions in personal computers to ensure easier use has led to the development of more powerful chipsets. **Work problem** 工作問題 However, this trend increases the complexity of designing hardware and the difficulty of programming software such as in the BIOS, driver, and operating system. **Quantitative specification of problem** 問題的量化 The CMOS ROM capability in BIOS code is only 256k. Therefore, more code must be added to increase the number of functions. **Importance of problem** 問題的中心 Such an addition not only lowers CMOS capability and wastes motherboard space, but also increases production costs since a a larger CMOS ROM is required.

*Quality Control* 品保相關分類

**Setting of work proposal** 工作提案建構 Some IC designs have become commercially available without the entire verification process being executed, thus reducing the time to market. **Work problem** 工作問題 However, this approach creates potentially large financial risks for an IC design firm. **Quantitative specification of problem** 問題的量化 For instance, in 1995, Intel spent nearly 500 million US dollars in recalling Pentium CPUs that contained one floating-point division bug. **Importance of problem** 問題的中心 Unfortunately, the time required for the verification process could increase exponentially as IC designs become more complex.

*Electrical Engineering* 電子相關分類

**Setting of work proposal** 工作提案建構 Electrical parameters significantly affect wafer quality. Additionally, the acceptance criterion for these parameters is A2/R3, that is, three of five tested points must pass for a wafer to be accepted. **Work problem** 工作問題 Moreover, the inability to monitor online whether products are out of specification increases the likelihood of poor wafer quality, **Importance of problem** 問題的中心 endangering a company's competitiveness.

*Quality Control* 品保相關分類

**Setting of work proposal** 工作提案建構 OEM customers request that our company's chip sets receive accreditation from the Microsoft Windows Hardware Quality Labs to ensure hardware compatibility with Microsoft Windows operating systems. **Work problem** 工作問題 However, the IC design firm's lack of attention to both hardware (Chip-set) and software (Driver) prevents the firm from receiving accreditation from the Windows Hardware Quality Labs. **Importance of problem** 問題的中心 The inability to receive such accreditation makes our company's products less competitive.

*Computer Graphics* 電腦製圖相關分類

**Setting of work proposal** 工作提案建構 3D applications play an increasingly important role in daily living, especially in entertainment. **Work problem** 工作問題 However, improvements in software and hardware can not keep pace with consumers' preferences. **Importance of problem** 問題的中心 The inability to design a high performance graphic chip bounded with drivers severely limits 3D applications.

*Electrical Engineering* 電子相關分類

**Setting of work proposal** 工作提案建構 Hardware emulation is essential for verifying IC designs that are becoming larger and more complex. **Work**

**problem** 工作問題 However, emulators are often too expensive and slow. **Quantitative specification of problem** 問題的量化 A Quickturn emulator costs more than 1 million US dollars and only works below 1 MHz, which is markedly lower than the frequency required in newly developed devices. **Importance of problem** 問題的中心 Such a high cost prevents us from verifying the designs of platforms, and the low speed makes testing difficult and the results unrealistic.

*Electrical Engineering* 電子相關分類

**Setting of work proposal** 工作提案建構 Consumer demand for hard disc bandwidth has significantly increased in recent years owing to the large size of files and increased disc size. **Work problem** 工作問題 The conventional interface for hard discs cannot easily support such a high bandwidth. **Quantitative specification of problem** 問題的量化 The conventional interface requires many pin counts and the failure rate during manufacturing exceeds 5%, **Importance of problem** 問題的中心 making commercialization impossible.

*Chemical Engineering* 化工相關分類

**Setting of work proposal** 工作提案建構 Per Fluoride Compound (PFC) gases, including CF4, C2F6 and SF6, hurt the environment. **Work problem** 工作問題 Most industrial countries can not meet the aim of the Montreal Protocol goal to reduce the current usage of PFC gases by 10%.

**Quantitative specification of problem** 問題的量化 If sufficient information can not be gathered and the quantity of PFC gases used by factories can not be measured annually, **Importance of problem** 問題的中心 then the inability to decrease the use of PFC gases will accelerate the depletion of the ozone layer.

*Chemical Engineering* 化工相關分類

**Setting of work proposal** 工作提案建構 Beyond adhering to governmental air quality codes, a factory must impose strict environmental standards out of concern for the neighboring community. **Work problem** 工作問題 Nevertheless, exhaust in the form of visible, white smoke and an odor is occasionally emitted from factories. **Importance of problem** 問題的中心 The inability to reduce the visible, white smoke and the odor to acceptable levels not only creates a potentially unsafe working environment, but also poses a perceived threat to the nearby community.

*Electrical Engineering* 電子相關分類

**Setting of work proposal** 工作提案建構 High speed is an increasingly important feature of the 3D VGA Card in the PC industry. **Work problem** 工作問題 However, in a high speed PC, the resistor impedance, crosstalk, power quality, and EMI constitute bottlenecks for further development. **Quantitative specification of problem** 問題的量化 The inability to resolve these problems lowers the speed and overall performance of PCs.

**Importance of problem** 問題的中心 Additionally, a reduction in power quality causes the card to perform unsatisfactorily. Moreover, resistor impedance and cross-talk cause noise in the PCB, negatively impacting the VGA Card's speed during operation and falling below FCC standards with respect to EMI.

*Quality Control* 品保相關分類

**Setting of work proposal** 工作提案建構 Semiconductor processes are increasingly complex. **Work problem** 工作問題 Effectively monitoring the process stability of each module and accurately determining the results in a timely manner are increasingly difficult. **Quantitative specification of problem** 問題的量化 The inability to effectively monitor the process stability of each module and accurately determine the monitoring results in a timely manner **Importance of problem** 問題的中心 makes it impossible for many automation engineers to feedback in-line monitor data in real time.

*Chemical Engineering* 化工相關分類

**Setting of work proposal** 工作提案建構 Owing to environmental concerns, NF3 gas is gradually replacing CxFy gas in the dielectric CVD clean process. **Work problem** 工作問題 However, the global shortage of NF3 clean gas explains why it costs significantly more than CxFy clean gas, especially for high impurity ( > 4N ) NF3 clean gas. **Quantitative**

**specification of problem** 問題的量化 Replacing CxFy clean gas costs more than three times more than using high impurity NF3 clean gas. **Importance of problem** 問題的中心 Used in our fab's DCVD clean process for quite some time, high impurity NF3 clean gas is an expensive clean gas that creates a high overhead cost for our products.

*Quality Control* 品保相關分類

**Setting of work proposal** 工作提案建構 Capacity planning is essential in evaluating a wafer fab's capacity. For instance, a wafer fab's capacity is forecasted six months ahead to measure its productivity. **Work problem** 工作問題 However, conventional analysis models can not accurately forecast capacity over the short (dynamic) and long (static) term owing to the varying complexity of the model(s) employed. **Importance of problem** 問題的中心 The inability to accurately forecast capacity over the short (dynamic) and long (static) term will ultimately lower a company's competitiveness.

*Electrical Engineering* 電子相關分類

**Setting of work proposal** 工作提案建構 Approaching the shallow junction in ion implantation is complicated since the ion beam must be generated at an extremely low energy. **Work problem** 工作問題 Additionally, a high beam current can not be obtained when a low energy ion beam is generated, resulting in a low throughput. **Quantitative**

**specification of problem** 問題的量化 For example, if the condition, B+11/2keV/1e15, is desired, the machine can only generate a 1 mA beam current, **Importance of problem** 問題的中心 and thus requires an excessive time of 1 hour to finish one lot of ion implantation.

*Computer Science* 資訊科學相關分類

**Setting of work proposal** 工作提案建構 Evolution of global Internet commerce has led to improved solutions to problems of computer processes and Internet security. **Work problem** 工作問題 Conventional recognition approaches can not ensure individual security when commercial transactions are conducted on the Internet **Quantitative specification of problem** 問題的量化, as evidenced by statistics from the National Internet Institute that the Internet crime rate has increased by about three times in 2001. **Importance of problem** 問題的中心 The inability to ensure individual security during commercial, on-line transactions will negatively impact the global growth of this sector.

*Electrical Engineering* 電子相關分類

**Setting of work proposal** 工作提案建構 The micron-level semiconductor process is characterized by a limited number of transistors and the occupied area. Compared with the intrinsic delay of cells, the net delay is insignificant enough to be overlooked. In contrast, the deep sub-micron design involves cells with enhanced performance and a slight delay.

Therefore, net delay rather than cell delay is becoming the dominating factor in the semiconductor process. **Work problem** 工作問題 Unfortunately, the wireload model utilized in synthesis provides inadequate information regarding the routed net because it only estimates the delays from the fanout number and possible routing path. **Quantitative specification of problem** 問題的量化 The inaccuracy of wireload models' determination of wire delays ranges from roughly 10% to 30%. **Importance of problem** 問題的中心 Neglecting the assigned locations of the cells during synthesis makes it impossible to evaluate the wire delays and achieve optimization efficiently.

*Electrical Engineering* 電子相關分類

**Setting of work proposal** 工作提案建構 Consumer demand for high speed transmission between processors and peripheral storage devices in computer systems has significantly increased in recent years. **Work problem** 工作問題 However, conventional parallel bus architecture has a relatively low transmission speed owing to interference between bus lines. **Quantitative specification of problem** 問題的量化 For instance, while the parallel bus hard-disk can only operate at 100 MB/sec, the serial one can operate over a range of 150 MB/sec to 600 MB/sec. **Importance of problem** 問題的中心 Consequently, the lack of a high speed storage device decreases the efficiency of the entire computer system.

*Quality Control* 品保相關分類

**Setting of work proposal** 工作提案建構 An increasing number of interest rate derivative products priced according to the forward interest rate have highlighted the empirical challenges of fitting forward rate yield curves to current market data. **Work problem** 工作問題 However, conventional methods, which only focus on fitting a yield curve and transforming it into a forward rate curve, result in an unreasonable, extremely high or negative forward rate, **Quantitative specification of problem** 問題的量化 such as 150% or -60%, respectively. **Importance of problem** 問題的中心 Pricing interest rate derivatives by adopting these inappropriate fitting forward rate methods will result in large deviations from market data.

*Quality Control* 品保相關分類

**Setting of work proposal** 工作提案建構 The amount of paper used in offices is rapidly and globally falling. For instance, many companies use their own intranets and document management systems to control the circulation of security documents effectively. **Work problem** 工作問題 However, encouraging employees to view materials on-line instead of on paper is difficult. **Importance of problem** 問題的中心 Additionally, paper usage involves too much time in acquiring necessary signatures for a particular document.

*Electrical Engineering* 電子相關分類

**Setting of work proposal** 工作提案建構 High performance IC package design has become increasingly complex in current applications. **Work problem** 工作問題 However, the conventional notion of simply "packing" the IC can not satisfy current requirements. **Importance of problem** 問題的中心 For instance, IC package design that does not consider high performance, low cost and reliability will lead to a fall in an IC design company's competitiveness.

*Electrical Engineering* 電子相關分類

**Setting of work proposal** 工作提案建構 Making a chipset involves using many advanced technologies in circuit design and wafer fabrication, as evidenced by the dramatic increase in speed and performance of personal computers. **Work problem** 工作問題 However, the increasingly compact size of wafer chipsets causes problems in design and process. **Quantitative specification of problem** 問題的量化 Although increasing the wafer size and reducing the device's dimensions are strongly desired, a deep submicron process, 0.15um or lower, **Importance of problem** 問題的中心 will decrease not only the product yield and quality, but also a company's technical and market position. Therefore, in determining the appropriate design and process, IC designers must adopt effective wafer design processes to increase product yield and lower overhead costs.

*Computer Science* 資訊科學相關分類

**Setting of work proposal** 工作提案建構 When surfing the Web using a typical modem, users have become accustomed to heavy network congestion on the Internet highway that slows Internet access. The recently introduced ADSL (Asymmetric Digital Subscriber Line) technology may ease frustration among users. **Work problem** 工作問題 However, the codeword size is equal to 255, and the depth can reach 64 in the interleaving block of the ADSL chip. **Importance of problem** 問題的中心 Accordingly, the chip has a large die area owing to a memory requirement of up to 16K bytes.

*Electrical Engineering* 電子相關分類

**Setting of work proposal** 工作提案建構 Hardware emulation is essential for verifying IC designs that are becoming larger and more complex. **Work problem** 工作問題 However, emulators, for example, our emulation solution, are too expensive and too slow. **Quantitative specification of problem** 問題的量化 A Quickturn emulator costs more than a million US dollars and only works below 1 MHz, which is markedly lower than the frequency required in newly developed devices. **Importance of problem** 問題的中心 Such a high cost prevents us from verifying the designs of platforms, and the low speed makes testing difficult and the results unrealistic.

*Electrical Engineering* 電子相關分類

**Setting of work proposal** 工作提案建構 Despite the obstacles to accurately forecasting capacity planning, our company strives to enhance its market competitiveness by developing a more precise model. **Work problem** 工作問題 Although a viable solution to this problem, the dynamic capacity model requires a tremendous amount of data input. Generally, more data input implies a more accurate model. Restated, an accurate capacity forecast depends on sufficient input. According to our estimates, the dynamic capacity forecast and the actual throughput diverge by less than 10%. **Quantitative specification of problem** 問題的量化 Given the necessity of a stable server, a situation in which the server is down more than twice monthly will negatively impact our data reliability. **Importance of problem** 問題的中心 As well known, the CIM system is occasionally unstable. Subsequent deviation may require close collaboration between departments.

# *Unit Two*

Tom     Industrial Engineering(工業工程)

Susan     Information Management(資訊管理)

# Outlining the work proposal (part two): describing the plan to solve the problem

## 工作提案計畫(第二部分):行動

**Vocabulary and related expressions**   相關字詞

efficient evaluation mode   有能力的評價模式
brands   廠牌
evaluate the cost and effectiveness   評價成本和效能
viable   可實行的
a valuable reference   有價值的參考
GIS-based architecture   以地理資訊系統為基礎的架構
handheld mobile devices   手持通訊設備
automatically page   自動通知
PDA users   PDA使用者
wireless network   無線的網路
access information   取得資訊
filtered   過濾器
networked peer assessment system   網路上能力評價系統
higher education   高等教育
students' attitudes   學生的態度
reliability and validity coefficients   可靠與有效係數
feasibility   可行性
alternative strategy   替代的方法
distance learning   遠距教學

**A** Write down the key points of the situations on the preceding page while the instructor reads aloud the script on page 324.

**Situation 1**

_____

_____

_____

**Situation 2**

_____

_____

_____

**Situation 3**

_____

_____

_____

**B** Based on the three situations in this unit, write three questions beginning with *How*, and answer them.

**Examples**

*How can one use Tom 's evaluation model?*

*To evaluate the cost and effectiveness of all viable bus systems, and provide government with a valuable reference for selecting such systems*

*How can Susan 's GIS-based architecture automatically page PDA users?*

*Through a wireless network*

1. _____

    _____

    _____

2. _____

    _____

    _____

3.

_____

_____

**C** Based on the three situations in this unit, write three questions beginning with **Why**, and answer them.

**Examples**

*Why does John want to develop a networked peer assessment system capable of supporting instruction and learning?*

*To analyze the effectiveness of students' learning in higher education.*

*Why is Susan's GIS-based architecture unique?*

*It supports a service that automatically reports to handheld mobile devices.*

1. _____

_____

2 _____

_____

3.

_____

_____

D Based on the three situations in this unit, write three questions beginning with **What**, and answer them.

**Examples**

*What does Susan want to do?*

*She wants to design a GIS-based architecture that supports a service that automatically reports to handheld mobile devices.*

*What is John's networked peer assessment system capable of doing?*

*Supporting instruction and learning to analyze the effectiveness of students' learning in higher education*

1.

_____

_____

2.

_____

_____

3.

_____

_____

## E Write questions that match the answers provided.

**Examples**

*How can one use John's efficient evaluation model?*

*To select brands of buses which run on natural gas*

*Who can be provided with a valuable reference for selecting systems that run on natural gas?*

*The government*

*Unit* Outlining the work proposal (part two): describing the plan to
*Two* solve the problem
工作提案計畫（第二部分）：行動

1.

_____

_____

PDA users

2.

_____

_____

A GIS-based architecture that supports a service that automatically reports to handheld mobile devices

3.

_____

_____

Cost and effectiveness

# Unit Two

Outlining the work proposal (part two):
describing the plan to solve the problem

工作提案計畫（第二部分）：行動

1 Work objective 工作目標
2 Anticipated results 希望的結果
3 Contribution to field 領域的貢獻

Outlining the second part of the work proposal involves describing the plan to solve the problem. The plan consists of the following parts.

**1. Work objective 工作目標. What is the objective of your work proposal? 工作提案的目標？**

Consider the following examples.

◎develop a method of converting a 3D display to a stereo 3D display, using compatible hardware and software without special hardware.

◎develop an HDP-CVD process capable of reducing the Ar gas flow from 390 sccm to 50 sccm and controlling the pressure from 5 mtorr to 2.5 mtorr.

◎design a numerical method to choose the bandwidth of the loop filter efficiently and an additional apparatus to estimate offsets precisely.

◎develop a verification process capable of ensuring the accuracy of an IC design in a relatively short time

◎construct an on-line SPC system capable of detecting OOS (out of specification) points and displaying the message by e-mail to the owner on time

◎develop an analytical geometry method capable of accurately predicting the geometric parameters of a C4 type solder joint in flip chip technology after a reflow process

◎design an effective GAs model capable of deriving an optimal solution for each state in a multi-state problem

◎develop an estimation method capable of obtaining the marginal number of order parameter dimensions

◎elucidate the critical behavior of a binary symmetrical mixture using the collective variables method

◎attempt to confirm the defect-recombination mechanism as appropriate for exoemission from CsBr -similar structures

◎develop a novel order-disorder four-state model capable of theoretically describing the dielectric properties of DMAGaS-DMAAIS ferroelectric crystals

◎investigate the role of RSOE in II-VI semiconductors and construct a related model

◎calculate the theoretical exoelectronic energy spectra to determine whether the defect-recombination mechanism is responsible for EEE from CsBr.

◎study how scattering centers influence the electrical conductivity of metal films using a novel evaporation method

◎investigate the formation of radiation-stimulated defects in doped cesium iodide crystals

◎develop a novel competitiveness model capable of integrating a global optimization algorithm into the IMD World Competitiveness Model

**2. Anticipated results 希望的結果. What are the main results that you hope to achieve in your project? 你希望達成的結果?**

Consider the following examples.

◎In addition to adhering to FCC specifications, the proposed 3D VGA board can achieve a higher frequency and fewer coupling problems than conventional boards.

◎The proposed system can automatically package products and test those products over one to three days without user intervention. The proposed system can also store test results on a log file for further examination.

◎The proposed design can lock a wide range of offsets at more than 100 KHz in a short acquisition time with a symbol error rate of less than 0.01.

◎The proposed strategy can diminish the visibility of factory-emitted exhaust from current levels of 20% to 0% and remove the odor by upgrading air treatment equipment.

◎The proposed algorithm can integrate manual routines into a single procedure, reducing operational time by approximately 10% to 30%.

◎The proposed method can estimate the location and dispersion effects of an optimization procedure more accurately than conventional methods when solving problems related to ordered categorical data quality.

◎The proposed model can predict the residual level of groundwater contaminants to an accuracy of 95%.

◎The proposed algorithm can reduce the computational time required to solve a nonlinear programming model by 50% of that required by piecewise linearization algorithms.

◎The proposed model can reduce the time to construct a 3D face by 10%, by optimizing the 3D model rather than manually retrieving 2D images.

◎The proposed system can identify 85% of all network problems in 5 minutes by reducing the complexity of operating a network management system.

◎The proposed method can estimate the marginal dimension of an order parameter, corresponding to experimental data found in pertinent literature.

◎The proposed investigation can quantify the non-universal critical characteristics of a hard-sphere square-well binary symmetrical mixture.

◎The proposed investigation can determine the effect of gamma-irradiation on the ionic conductivity of OH-doped cesium halides grown by different methods.

◎The proposed model can clarify the dielectric phenomena during the transition and disappearance of the ferroelectric phase under hydrostatic pressure.

◎The novel evaporation method can reduce the electrical resistance of thin films by 10 % when the thickness of metal films is constant.

**3. Contribution to field 領域的貢獻.  What is the main contribution of your project to the field or industrial sector to which your company or organization belongs? 你的提案對相關工作領域的貢獻?**

Consider the following examples.

◎The proposed test parameters can be used to establish a productive qualification flow, thereby upgrading capacity and reliability.

◎A digital receiver that adopts the proposed method during carrier recovery can outperform conventional models in terms of tracking, making telecommunication products more competitive.

◎The proposed driver bounded with our graphics chip can elevate the company's technical and marketing position within this area of product development.

◎The proposed analytic method can enable our company to forecast more accurately short term and long term productivity, enhancing our market competitiveness.

◎By omitting redundant processing steps, the proposed algorithm can reduce the overall design cycle time by more than one third.

◎Our results can provide a valuable reference for governmental authorities when attempting to remediate contaminants.

◎The easy-to-use editor can be used on-line for mathematics courses, helping users to more easily edit complex mathematical symbols.

◎The networked electronic portfolio system and peer assessment can be used at all educational levels.

◎The proposed algorithm can obtain the global optimum in general nonlinear programming models to within a tolerable error and significantly increase computational efficiency by decreasing the use of 0-1 variables.

◎The proposed model can minimize the tolerable errors associated with constructing a digital face, enhancing multimedia or animation applications by reducing formulation costs and creating more realistic digital objects.

◎The proposed system allows for the design of flexible and inexpensive network management for enterprises by combining free software from the Internet.

◎The proposed learning environment can eliminate the need for hand coding GA programs, thus simplifying the process of learning genetic algorithms for GA students.

◎The proposed investigation can calculate how thermodynamic functions depend on microscopic parameters near binary symmetrical mixture critical points, clarifying how microscopic parameters influence macroscopic critical behavior.

◎The proposed investigation can clarify the EEE mechanism for CsBr, shedding further light on the formation of exoelectrons.

◎The novel evaporation method can eliminate the structure of scattering centers, providing further insight into the structure of thin films and scattering centers.

◎By applying the method of ionic current of thermodepolarization (ICT) to examine CsJ crystals, the proposed investigation can provide further insight into dipole relaxation processes in doped CsJ crystals.

**F** In the space below, outline the second part of your work proposal by describing the plan to solve the problem.

Work objective 工作目標. Based on the above, we should

_____

_____

_____

Anticipated results 希望的結果:

_____

_____

_____

Contribution to field 領域的貢獻:

_____

_____

_____

**G** Look at the following examples of how to outline the second part of your work proposal by describing the plan to solve the problem.

*Civil Engineering* 土木相關分類

**Work objective** 工作目標 Based on the above, we should develop an optimal operating model capable of choosing an effective monitor to adequately control a groundwater system. **Anticipated results** 希望的結果 The proposed model can reduce much of the cost and time associated with drilling a new well. **Contribution to field** 領域的貢獻 If cost and time can be reduced by 10% of those of conventional models, the proposed model can significantly contribute to the design of a groundwater supply network.

*Industrial Engineering* 工業工程相關分類

**Work objective** 工作目標 Based on the above, we should develop a process capability index that adopts Clement's and Bootstrap methods to overcome such obstacles. **Anticipated results** 希望的結果 Using Clement's method, the process capability index proposed herein can evaluate a non-normal distribution process using conventional indices. Additionally, by applying the Bootstrap confidence interval estimator, the proposed process capability index can reduce the sampling error. **Contribution to field** 領域的貢獻 Importantly, the proposed process capability index can allow engineers without a statistical background to make decisions easily and quickly when comparing manufacturing processes or selecting an alternative supplier.

*Civil Engineering* 土木相關分類

**Work objective** 工作目標 Based on the above, we should design an optimization model capable of effectively managing water resources of current reservoirs to decrease the probability of water shortages. **Anticipated results** 希望的結果 The proposed model can accurately reflect how multi-objectives compete with each other and estimate the available releases of multi-reservoirs, in contrast to conventional  water resource management models that can not simultaneously do both. **Contribution to field** 領域的貢獻 Thus, the proposed optimization model can provide a valuable reference for governmental authorities when drawing up water resource-related management strategies.

*Industrial Engineering* 工業工程相關分類

**Work objective** 工作目標 Based on the above, we should construct an efficient quality control process capable of detecting assignable causes concealed behind both multiple characteristics and multiple readings in a manufacturing system with several sources of variation. **Anticipated results** 希望的結果 The quality control process can markedly reduce the number of false alarms due to assignable causes, saving considerable manpower and capital. In addition, the quality control process proposed herein can appropriately detect small shifts in production phase and identify different sources of variation. **Contribution to field** 領域的貢獻 Accordingly, an engineer can rapidly adjust the manufacturing system to solve related problems and enhance wafer quality.

*Civil Engineering* 土木相關分類

**Work objective** 工作目標 Based on the above, we should develop a NAPL simulator model that includes several parameters acquired experimentally. **Anticipated results** 希望的結果 The proposed model can predict the residual level of groundwater contaminants to an accuracy of

95%, **Contribution to field** 領域的貢獻 providing a valuable reference for governmental authorities when attempting to remediate contaminants.

*Civil Engineering* 土木相關分類

**Work objective** 工作目標 Based on the above, we should develop a deterministic and stochastic model for simulating groundwater flow to assess monitoring network alternatives. **Anticipated results** 希望的結果 The proposed stochastic model can provide further insight into the uncertainty of the estimation error by adopting different groundwater monitoring network alternatives. **Contribution to field** 領域的貢獻 Additionally, the proposed model can minimize the costs of constructing a monitoring network by assessing monitoring network alternatives. Results in this study can be used to construct a real-time groundwater flow model for supporting conjunctive use of water resources by combining the stochastic model and real-time groundwater level measurements.

*Industrial Engineering* 工業工程相關分類

**Work objective** 工作目標 Based on the above, we should develop a flexible and accurate neural network structure that applies artificial intelligence and fuzzy theory to business ratings and bankruptcy prediction. **Anticipated results** 希望的結果 The proposed model can discriminate multiple ranks with 90% accuracy. **Contribution to field** 領域的貢獻 By incorporating subjective judgment, the neural network structure proposed herein can allow an organization to assess the degree of risk and whether an enterprise will become insolvent.

*Statistics* 統計相關分類

**Work objective** 工作目標 Based on the above, we should develop an efficient response surface method capable of optimizing ordered categorical data process parameters since continuously setting the process parameters leads to optimization of the location and dispersion effects. **Anticipated results** 希望的結果 The proposed method can estimate the location effect and dispersion effect of an optimization procedure more accurately than conventional methods when solving problems related to ordered categorical data quality. **Contribution to field** 領域的貢獻 Additionally, the proposed method includes an optimization procedure with simplified calculations of ordered categorical data for engineers.

*Industrial Engineering* 工業工程相關分類

**Work objective** 工作目標 Based on the above, we should develop an efficient hypothesis testing procedure for PCIS, capable of assessing the operational cycle time (OCT) and delivery time (DT) for VLSI. **Anticipated results** 希望的結果 The proposed procedure can monitor the OCT and DT to enhance the competitiveness of suppliers. **Contribution to field** 領域的貢獻 The proposed procedure can be used to construct a testing procedure for PCIs to assess OCT and DT for suppliers.

*Distance Learning* 遠距教學相關分類

**Work objective** 工作目標 Based on the above, we should develop an Internet-based constructive learning environment that allows participants to interactively link their conceptual maps for accumulative learning. **Anticipated results** 希望的結果 The proposed learning environment can prevent unequal participation in group activities. **Contribution to field** 領域的貢獻 Additionally, the proposed learning environment can provide a convenient cooperative-competitive model for educators to use in classroom activities.

*Civil Engineering* 土木相關分類

**Work objective** 工作目標 Based on the above, we should develop an NAPL simulator model that includes several parameters acquired experimentally. **Anticipated results** 希望的結果 The proposed model can predict the residual level of groundwater contaminants to an accuracy of 95%, **Contribution to field** 領域的貢獻 providing a valuable reference for governmental authorities when attempting to remediate contaminants.

*Materials Science* 材料相關分類

**Work objective** 工作目標 Based on the above, we should develop a processing method capable of producing materials of a required shape at a relatively low cost, directly from wrought products. **Anticipated results** 希

望的結果 The proposed processing method can increase the elongation of magnesium alloys by over 20% through significantly reducing costs and time. **Contribution to field** 領域的貢獻 Thus, the proposed processing method can significantly contribute to the role of magnesium alloys in the market for structural materials.

*Computer Science* 資訊科學相關分類

**Work objective** 工作目標 Based on the above, we should develop a mathematical editor using a Java applet on the Web, capable of using a graphic user interface to edit mathematical symbols. **Anticipated results** 希望的結果 The proposed editor can allow complex equations to be edited on the Web through a graphic user interface. **Contribution to field** 領域的貢獻 The easy-to-use editor can be used on-line in mathematical courses, helping users to edit complex mathematical symbols.

*Distance Learning* 遠距教學相關分類

**Work objective** 工作目標 Based on the above, we should develop a networked electronic portfolio system with peer assessment to provide a creative means of assessing the higher-order thinking of students. **Anticipated results** 希望的結果 The proposed system can allow teachers to observe the students' higher-level thinking by collecting records of students' homework and interaction with peers on the networked electronic portfolio system via peer assessment. **Contribution to field** 領域的貢獻

Additionally, the networked electronic portfolio system and peer assessment can be used at all educational levels.

*Distance Learning* 遠距教學相關分類

**Work objective** 工作目標 Based on the above, we should analyze the online discussions of collaborative teams to reveal whether such activity engages students in higher order thinking and how it takes place. **Anticipated results** 希望的結果 The proposed analysis can identify what, how, and where higher order thinking is embedded in the online discussions of collaborative design teams. **Contribution to field** 領域的貢獻 Additionally, the proposed analysis can describe more specifically the merits of students' collaborative design through online discussion.

*Computer Science* 資訊科學相關分類

**Work objective** 工作目標 Based on the above, we should develop an enhanced piecewise linearization algorithm, capable of obtaining the global optimum of a nonlinear model, for use in a web-based optimization system. **Anticipated results** 希望的結果 The proposed algorithm can reduce the computational time required to solve a nonlinear programming model by 50% of that required by piecewise linearization algorithms. **Contribution to field** 領域的貢獻 Additionally, the proposed algorithm can obtain the global optimum in general nonlinear programming models within a tolerable error and significantly increase computational efficiency by decreasing the use of 0-1 variables.

*Information Management* 資訊管理相關分類

**Work objective** 工作目標 Based on the above, we should develop an efficient face model capable of formulating the 3D image of an individual's face from three 2D images. **Anticipated results** 希望的結果 The proposed model can reduce the time required to construct a 3D face by 10%, by optimizing the 3D model rather than manually retrieving 2D images. **Contribution to field** 領域的貢獻 Additionally, the proposed model can minimize the tolerable errors associated with constructing a digital face, thus enhancing multimedia or animation applications by reducing formulation costs and creating more realistic digital objects.

*Information Management* 資訊管理相關分類

**Work objective** 工作目標 Based on the above, we should develop a novel web-based system in a PC-LAN environment capable of detecting network problems. **Anticipated results** 希望的結果 The proposed system can identify 85% of all network problems in 5 minutes by reducing the complexity of operating a network management system. **Contribution to field** 領域的貢獻 Additionally, the proposed system can allow for the design of flexible and inexpensive network management for enterprises by combining free software from the Internet.

70

*Computer Science* 資訊科學相關分類

**Work objective** 工作目標 Based on the above, we should develop a novel learning environment capable of assisting students in learning genetic algorithms, supported flexibly by computer-assisted instruction. **Anticipated results** 希望的結果 The proposed environment can reduce the time required for students to complete a GA assignment to one week, increasing the number of practice exercises that can be implemented and allowing them to better learn GAs. **Contribution to field** 領域的貢獻 Additionally, the proposed learning environment can eliminate the need for hand coding of GA programs, simplifying the process of learning genetic algorithms.

*Mechanical Engineering* 機械相關分類

**Work objective** 工作目標 Based on the above, we should develop an analytical geometry method capable of accurately predicting the geometric parameters of a C4 type solder joint in flip chip technology after a reflow process. **Anticipated results** 希望的結果 The proposed method can predict geometric parameters of a C4 type solder joint to within 5% of those obtained by a specific method found in the literature. **Contribution to field** 領域的貢獻 Additionally, the proposed method can be used to design the geometric parameters of a C4 type solder joint, and can enhance the reliability of a flip chip package and reduce its stress concentration.

*Computer Science* 資訊科學相關分類

**Work objective** 工作目標 Based on the above, we should design an effective GA model capable of deriving an optimal solution for each state in a multi-state problem. **Anticipated results** 希望的結果 The proposed model can increase the accuracy of an optimum solution derived for a particular multi-state problem. **Contribution to field** 領域的貢獻 Additionally, the proposed model can enhance conventional genetic algorithms by systematically solving multi-state problems through the use of the polyploidy concept.

*Physics* 物理相關分類

**Work objective** 工作目標 Based on the above, we should develop an estimation method capable of obtaining the marginal number of order parameter dimensions. **Anticipated results** 希望的結果 The proposed method can estimate the marginal dimension of an order parameter, corresponding to experimental data found in pertinent literature. **Contribution to field** 領域的貢獻 Moreover, the method can be used to apply several theoretical renormalization group methods, in addition to those available in the literature, yielding accurate results for a weakly diluted O (m) model that can thoroughly describe critical behavior.

*Physics* 物理相關分類

**Work objective** 工作目標 Based on the above, we should elucidate the critical behavior of a binary symmetrical mixture using the collective variables method. **Anticipated results** 希望的結果 The proposed investigation can quantify the non-universal critical characteristics of a hard-sphere square-well binary symmetrical mixture. **Contribution to field** 領域的貢獻 Additionally, the proposed investigation can calculate how thermodynamic functions depend on microscopic parameters near binary symmetrical mixture critical points, clarifying how microscopic parameters influence macroscopic critical behavior.

*Physics* 物理相關分類

**Work objective** 工作目標 Based on the above, we should attempt to confirm the defect- recombination mechanism as appropriate for exoemission from CsBr -similar structures. **Anticipated results** 希望的結果 The proposed investigation can determine the effect of gamma-irradiation on the ionic conductivity of OH-doped cesium halides grown by different methods. **Contribution to field** 領域的貢獻 Additionally, the proposed investigation can distinguish between OH-doped and undoped Cs-halides in terms of radiation defect rebuilding, and estimate the radiation stability of cesium halides.

*Physics* 物理相關分類

**Work objective** 工作目標 Based on the above, we should develop a novel order-disorder four-state model capable of theoretically describing the dielectric properties of DMAGaS-DMAAlS ferroelectric crystals. **Anticipated results** 希望的結果 The proposed model can clarify the dielectric phenomena during the transition and disappearance of the ferroelectric phase under hydrostatic pressure. **Contribution to field** 領域 的貢獻 Additionally, the proposed model can accurately describe systems with more than two equilibrium positions on a site, extending conventional order-disorder models to a wider class of materials.

*Physics* 物理相關分類

**Work objective** 工作目標 Based on the above, we should investigate the role of RSOE in II-VI semiconductors and construct a related model. **Anticipated results** 希望的結果 The proposed model can clarify low-dose radiation processes in solids and improve the parameters of structures based on II-VI semiconductors. **Contribution to field** 領域的貢獻 Additionally, the proposed model can be used not only to increase the range/number of objects in which the RSOE is observed, but also to distinguish between II-VI and other semiconductors, (Si, III-V semiconductors) in terms of the RSOE mechanism.

*Physics* 物理相關分類

**Work objective**工作目標 Based on the above, we should calculate the theoretical exoelectronic energy spectra to determine whether the defect-recombination mechanism is responsible for EEE from CsBr. **Anticipated results** 希望的結果 The proposed investigation can allow the energy spectra of exoelectrons to be obtained from CsBr. **Contribution to field** 領域的貢獻 Additionally, the proposed investigation can clarify the EEE mechanism for CsBr, shedding further light on the formation of exoelectrons .

*Physics* 物理相關分類

**Work objective** 工作目標 Based on the above, we should study how scattering centers influence the electrical conductivity of metal films, using a novel evaporation method. **Anticipated results** 希望的結果 The novel evaporation method can reduce the electrical resistance of thin films by 10 % when the thickness of metal films is constant. **Contribution to field** 領域的貢獻 Additionally, the novel evaporation method can eliminate the structure of scattering centers, providing further insight into the structure of thin films and scattering centers.

*Physics* 物理相關分類

**Work objective** 工作目標 Based on the above, we should investigate the formation of radiation-stimulated defects in doped cesium iodide crystals. **Anticipated results** 希望的結果 The proposed investigation can determine the effect of gamma irradiation on defect formation in CsJ crystals doped with different impurities. **Contribution to field** 領域的貢獻 By applying the method of ionic current of thermodepolarization (ICT) to examine CsJ crystals, the proposed investigation can provide further insight into the dipole relaxation processes in doped CsJ crystals.

*Information Management* 資訊管理相關分類

**Work objective** 工作目標 Based on the above, we should develop a novel competitiveness model capable of integrating a global optimization algorithm into the IMD World Competitiveness Model. **Anticipated results** 希望的結果 The proposed model can increase the availability of the IMD World Competitiveness Model by 20%. **Contribution to field** 領域的貢獻 In doing so, the proposed method can bolster Taiwan's world competitiveness ranking to within the top 10.

*Computer Science* 資訊科學相關分類

**Work objective** 工作目標 Based on the above, we should develop a method of converting a 3D display to a stereo 3D display, using compatible hardware and software without special hardware. **Anticipated results** 希望的結果 As anticipated, the proposed method can generate real-time (60 frames per second), high-quality (1024x768, true color) and low-cost stereoscopic effects. **Contribution to field** 領域的貢獻 Upgrading to stereo 3D display can offer a more realistic 3D experience for users.

*Chemistry* 化學相關分類

**Work objective** 工作目標 Based on the above, we should develop an HDP-CVD process capable of reducing the Ar gas flow from 390 sccm to 50 sccm and controlling the pressure from 5 mtorr to 2.5 mtorr. **Anticipated results** 希望的結果 The proposed process can achieve an average aspect ratio of 2.7 for 0.2 um metal spacing. **Contribution to field** 領域的貢獻 The HDP-CVD process proposed herein can significantly enhance the gap filling capacity to satisfy 0.15 um process requirements.

*Quality Control* 品保相關分類

**Work objective** 工作目標 Based on the above, we should develop a program capable of creating a tariff. **Anticipated results** 希望的結果 In addition to displaying company information such as import and export figures, the proposed program can calculate various custom duties. **Contribution to field** 領域的貢獻 The proposed program can reduce the time to create a tariff because it only requires a minimal number of inputs to complete a task.

*Industrial Engineering* 工業工程相關分類

**Work objective** 工作目標 Based on the above, we should develop an accurate measurement method capable of assessing the capability of a measurement system. **Anticipated results** 希望的結果 The proposed method can estimate whether the measurement system is appropriate for use in analyzing a manufacturing process. **Contribution to field** 領域的貢獻 The proposed method can enhance the reliability of a manufacturing process by ensuring that the adopted measurement system is effective.

*Electrical Engineering* 電子相關分類

**Work objective** 工作目標 Based on the above, we should design a better 3D VGA board with less noise, capable of matching the resistor impedance as and reducing the EMI in the VGA board. **Anticipated results** 希望的結果 In addition to adhering to FCC specifications, the proposed 3D VGA board can achieve a higher frequency and fewer coupling problems than conventional boards . **Contribution to field** 領域的貢獻. The proposed 3D VGA board can reduce the adverse impact of the EMI so that all household electrical appliances that involve EMI can be used safely.

*Electrical Engineering* 電子相關分類

**Work objective** 工作目標 Based on the above, we should develop effective test parameters and demonstrate their effectiveness. **Anticipated results** 希望的結果 According to the proposed test parameters, more exact knowledge of the shortage of test items implies a shorter test time and more reliable product quality. **Contribution to field** 領域的貢獻 The proposed test parameters can be used to establish a productive qualification flow, upgrading capacity and reliability.

*Computer Science* 資訊科學相關分類

**Work objective** 工作目標 Based on the above, we should transfer the original simple database type to "SQL", which can support many records. **Anticipated results** 希望的結果 Following successful transfer, our database can record all drivers' testing problems for many years into the future. **Contribution to field** 領域的貢獻 In doing so, database tasks can be more easily executed; for example, the database can be automatically backed up daily or weekly.

*Electrical Engineering* 電子相關分類

**Work objective** 工作目標 Based on the above, we should develop an automotive testing system capable of overcoming such a shortage. **Anticipated results** 希望的結果 The proposed system can automatically package products and test those products in one to three days without user intervention. The proposed system can also store test results on a log file for further examination. **Contribution to field** 領域的貢獻 Testing can also be performed on a Saturday or Sunday to implement successfully the proposed system. Doing so can allow us to increase the number of tested items without requiring engineers to spend additional time in testing, thereby upgrading the overall development process.

*Electrical Engineering* 電子相關分類

**Work objective** 工作目標 Based on the above, we should design both a numerical method to choose efficiently the bandwidth of the loop filter and an additional apparatus to estimate offsets precisely. **Anticipated results** 希望的結果 The proposed design can lock a wide range of offsets of more than 100 KHz in a short acquisition time with a symbol error rate of less than 0.01. **Contribution to field** 領域的貢獻 A digital receiver that adopts the proposed method during carrier recovery can outperform conventional models in terms of tracking, making telecommunication products more competitive.

*Electrical Engineering* 電子相關分類

**Work objective** 工作目標 Based on the above, we should design an IC aging test system in which aging and burning tests can identify wafer defects. **Anticipated results** 希望的結果 The proposed IC aging test system can eliminate the need to use weak components in production. **Contribution to field** 領域的貢獻 The proposed IC aging test system can strengthen a company's overall market competitiveness by eliminating wafer defects and minimizing overhead costs.

*Electrical Engineering* 電子相關分類

**Work objective** 工作目標 Based on the above, we should develop a

precise logarithm device in a vertex shader. **Anticipated results** 希望的結果 The proposed logarithm device can lower costs below those of commercially available devices and  can operate at a frequency of 200 MHz. **Contribution to field** 領域的貢獻 A full-precision logarithm device can enhance the accuracy of the lighting effect and provide excellent graphics quality.

*Quality Control* 品保相關分類

**Work objective** 工作目標 Based on the above, we should develop a verification process capable of ensuring the accuracy of an IC design in a relatively short time. **Anticipated results** 希望的結果 The proposed verification process can focus on checking the functional boundary condition rather than the entire design, eliminating iterative testing of an identical function. **Contribution to field** 領域的貢獻 In addition to reducing the verification time, the proposed process can easily identify and resolve the founding error.

*Electrical Engineering* 電子相關分類

**Work objective** 工作目標 Based on the above, we should construct an on-line SPC system capable of detecting OOS (out of specification) points and displaying the message by e-mail to the owner on time. **Anticipated results** 希望的結果 The proposed system can enable an engineer to identify the OOS points and to determine whether that OOS belongs to a normal distribution. **Contribution to field** 領域的貢獻 In addition to reducing the disposal time, the proposed system can also maintain product quality.

*Computer Graphics* 電腦製圖相關分類

**Work objective** 工作目標 Based on above, we should develop a high quality 3D graphics driver with excellent performance. **Anticipated results** 希望的結果 The proposed 3D graphics driver can alleviate hardware limitations such as cost and die size trade-off, facilitating computer game development. **Contribution to field** 領域的貢獻 The proposed driver bounded with our graphics chip can elevate the company's technical and marketing position within this area of product development.

*Electrical Engineering* 電子相關分類

**Work objective** 工作目標 Based on the above, we should develop a serial interface that can easily support a high bandwidth throughput. **Anticipated results** 希望的結果 The proposed interface not only lowers the failure rate of manufacturing to 5% lower than that of the conventional interface, but also minimizes the pin counts to a fifth. **Contribution to field** 領域的貢獻 The new interface can make a high performance hard disc commercially feasible and can be easily upgraded in the future.

*Environmental Engineering* 環工相關分類

**Work objective** 工作目標 Based on the above, we should develop a

strategy to identify the origin of the factory-emitted exhaust in the form of visible, white smoke and an odor, and recommend how to abate its potential impact. **Anticipated results** 希望的結果 The proposed strategy can diminish the visibility of factory-emitted exhaust from current levels of 20% to 0% and remove the odor by upgrading air treatment equipment. **Contribution to field** 領域的貢獻 In addition to conforming to environmental codes, the proposed strategy can significantly reduce the perceived negative impact on the neighboring community.

*Quality Control* 品保相關分類

**Work objective** 工作目標 Based on the above, we should develop an analytic method that (regardless of the complexity of the model(s) employed) provides accurate capacity data to the Production Planning Department to optimally plan wafer production, and to the Marketing Department to fully realize the fab's constraints. **Anticipated results** 希望的結果 The proposed analytic method can provide accurate short-term capacity forecast data to the Production Planning Department for wafer production control and long-term capacity forecast data to the Marketing Department for strategy planning. **Contribution to field** 領域的貢獻 The proposed analytic method can enable our company to forecast more accurately short term and long term productivity, enhancing our market competitiveness.

*Electrical Engineering* 電子相關分類

**Work objective** 工作目標 Based on the above, we should design an effective method capable of reducingthe memory access channel bandwidth. **Anticipated results** 希望的結果 The proposed method can reduce the time to deliver data by 15%. **Contribution to field** 領域的貢獻 The proposed method can yield a high memory access channel bandwidth, increasing computational efficiency.

*Electrical Engineering* 電子相關分類

**Work objective** 工作目標 Based on the above, we should develop an algorithm capable of omitting redundant and time-consuming steps when applying design automation for wafer chips. **Anticipated results** 希望的結果 The proposed algorithm can integrate manual routines into a single procedure, reducing the operational time by approximately 10% to 30%. **Contribution to field** 領域的貢獻 By omitting redundant processing steps, the proposed algorithm can reduce the overall design cycle time by more than one third.

*Electrical Engineering* 電子相關分類

**Work objective** 工作目標 Based on above, we should develop a serial hard-disk architecture for an actual wafer product, capable of operating at high speeds. **Anticipated results** 希望的結果 The proposed architecture can ensure that the controller operates at 150 MB/sec for six months. **Contribution to field** 領域的貢獻 The proposed architecture, the first serial bus architecture of its kind, can elevate our company's technical and market position.

*Finance* 財務相關分類

**Work objective** 工作目標 Based on the above, we should develop a novel model to derive directly a reasonable forward rate curve in one explicit function to correlate well with market data. **Anticipated results** 希望的結果 The proposed model can include all observed market data on the yield curve and provide the smoothest possible forward rate curve consistent with the chosen functional form. **Contribution to field** 領域的貢獻 In contrast to conventional methods, the proposed model can provide an innovative means of directly fitting the forward rate with maximum smoothness and precision, providing a valuable solution for pricing interest rate derivatives.

*Quality Control* 品保相關分類

**Work objective** 工作目標 Based on the above, we should develop an effective strategy in which our company's technical documents are saved in E-file format, and Notes Client can be used to control all files. **Anticipated results** 希望的結果 The proposed strategy can economize on the use of paper in a company and reduces overhead costs. **Contribution to field** 領域的貢獻 Based on several tests of our company's circulation procedure

*Unit*
*Two*

Outlining the work proposal (part two): describing the plan to
solve the problem
工作提案計畫（第二部分）：行動

using Notes Client, the proposed strategy can offer an effective control
system ultimately to increase work productivity, and raise environmental
awareness among employees.

*Electrical Engineering* 電子相關分類

**Work objective** 工作目標 Based on the above, we should design an IC
product with excellent electrical / thermal performance, low cost, and high
reliability. **Anticipated results** 希望的結果 The proposed scheme, in its
mature stage of development, can help IC design engineers to minimize die
size, and achieve an appropriate package type, best package trace/lead
count fan-out, excellent thermal / electrical performance, and an optimized
design. **Contribution to field** 領域的貢獻 The proposed scheme can
shorten the design flow and dramatically reduce the time to market.

*Quality Control* 品保相關分類

**Work objective** 工作目標 Based on the above, we should develop a
strategy to solve design and production problems related to chipsets.
**Anticipated results** 希望的結果 The proposed strategy involves collecting
more production data, using the right equipment, seeking expert opinion,
and studying the latest technological developments. **Contribution to field**
領域的貢獻 The proposed strategy can allow our company to distinguish
its chipsets from the competition's, identify customer specifications,
prioritize work design and production objectives, and accurately forecast

the time to completion of production and the delivery time.

*Electrical Engineering* 電子相關分類

**Work objective** 工作目標 Based on the above, we should develop a strategy to understand the block that works in a chipset and to create more precise testing data. **Anticipated results** 希望的結果 The proposed strategy can allow testers to examine the test pattern more precisely than do conventional approaches. **Contribution to field** 領域的貢獻 The proposed strategy can reduce the inspection time by writing a rule-check program and familiarizing the user with related software.

*Electrical Engineering* 電子相關分類

**Work objective** 工作目標 Based on the above, we should develop an emulation method that costs less and performs better than the conventional one. **Anticipated results** 希望的結果 The proposed method can reduce the cost and time of developing at 10MHz by 95%, whereas the conventional method can only be used at a frequency of 1MHz. **Contribution to field** 領域的貢獻 The proposed method reduces testing time, includes new devices that can not function at 1MHz, and allows our company to verify the design for numerous platforms.

*Chemistry* 化學相關分類

**Work objective** 工作目標 Based on the above, we should develop a low impurity（3N）NF3 clean gas capable of reducing the costs of the DCVD cleaning process. **Anticipated results** 希望的結果 According to the clean test rate, the cleaning efficiency of the 3N NF3 gas should be comparable to that of the high impurity gas. According to the dielectric film's contamination test, the cleaning efficiency of the 3N NF3 gas should be comparable with that of both clean gases. However, the end-point time of the cleaning process should be the same for both clean gases. **Contribution to field** 領域的貢獻 The low impurity（3N）NF3 clean gas can reduce the costs of the DCVD cleaning process by more than 50%.

# Unit Three

 Jerry    Industrial Engineering(工業工程)

 Becky    Information Management(資訊管理)

## Writing the problem statement

問題描述

**Vocabulary and related expressions**    相關字詞

process quality    品質管制
delivery time    交貨時間
conventional process capability indices (PCIs)    慣例的製程能力指標
yield rate    產率
performance index    執行指標
objective procedure    客觀程序
inefficiency    無效率
high overhead cost    很高的成本
market competitiveness    市場競爭力
hypothesis testing procedure    假設檢定程序
assessing    評價
operational cycle time (OCT)    操作循環週期
geographical-based information    基礎地理資訊
passive mode of accessing information fails    使用不足資訊的消極方式
transmit geographic-based information    傳送基礎地理資訊
receive updated information in a timely manner    及時接收最新資訊
GIS-based architecture    以地理資訊系統為基礎的架構
Taiwan's global competitiveness ranking    台灣的全球競爭力排名
IMD World Competitiveness Scoreboard    洛桑管理學院世界競爭力排行榜
neglect    忽視
inaccurate decision    不正確的決定
integrating    整合
global optimization algorithm    全域最佳化演算法

Jerry — Industrial Engineering（工業工程）

Process quality
Delivery time
PCI
Process capability indices

Opponent
Market competitiveness

OCT New PCI
Operational cycle time
Delivery time

Becky — Information Management（資訊管理）

Effective ??

Timely manner

GIS-based architecture

Jack — Information Management（資訊管理）

World Competitiveness Scoreboard
IMD

Inaccurate decision

Competitiveness model

**A** Write down the key points of the situations on the preceding page while the instructor reads aloud the script on page 328.

**Situation 1**

_____

_____

_____

**Situation 2**

_____

_____

_____

**Situation 3**

_____

_____

_____

B Based on the three situations in this unit, write three questions beginning with **Why**, and answer them.

**Examples**

*Why are conventional process capability indices (PCIs) limited?*

*They can neither objectively assess quality and delivery time nor identify the relationship between PCIs and yield rate.*

*Why is Becky concerned with the passive mode of accessing information?*

*Because it fails to transmit geographic-based information toPDA users effectively*

1. _____

_____

2. _____

_____

3. _____

_____

C Based on the three situations in this unit, write three questions beginning with **How**, and answer them.

**Example**

*How will the inability to receive updated information in a timely manner affect PDA use?*

*It will severely limit the range of use of PDAs.*

1. _____

_____

2. _____

_____

3. _____

_____


**D** Based on the three situations in this unit, write three questions beginning with *What*, and answer them.

**Examples**

*What factor contributes to the fall in Taiwan's global competitiveness ranking in the IMD World Competitiveness Scoreboard?*

*The IMD Competitiveness Simulations neglect interactions among related factors.*

*What problem is Becky trying to solve?*

*The passive mode of accessing information fails to transmit effectively geographic-based information to PDA users.*

1. _____

_____

2. _____

_____

3. _____

_____

## E Write questions that match the answers provided.

**Examples**

*What can cause policy makers to make an inaccurate decision?*

*Use of the IMD Competitiveness Simulations*

*Who can benefit from geographic-based information in situations such as emergencies?*

*PDA users*

1. _____

   _____

   Firms that perform poorly in terms of quality and delivery

2. _____

   _____

   Inefficiency and a high overhead cost

3. _____

   _____

   Interactions among related factors

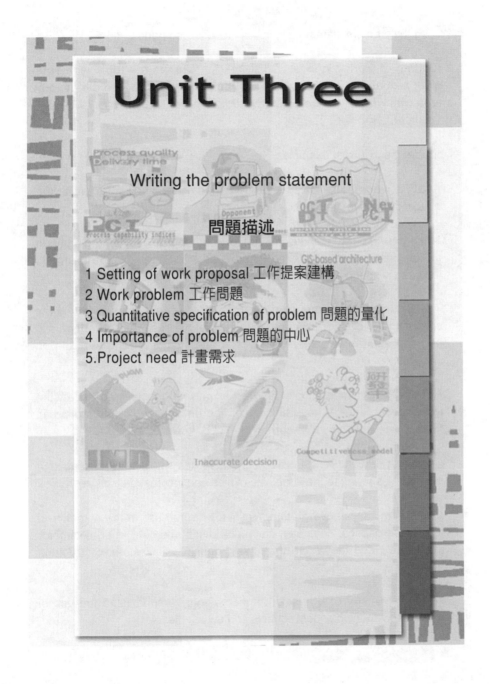

# Unit Three

## Writing the problem statement

## 問題描述

1 Setting of work proposal 工作提案建構
2 Work problem 工作問題
3 Quantitative specification of problem 問題的量化
4 Importance of problem 問題的中心
5.Project need 計畫需求

Writing the problem statement involves the following parts.

**1. Setting of work proposal 工作提案建構. What is the topic with your work proposal is concerned? Can you point the reader to the general area with which you are concerned?** 你工作提案的主題是什麼? 你的讀者可以明瞭工作提案的內容嗎?

Consider the following examples.

◎As IC designs become larger and increasingly complicated, hardware emulation is essential for their verification.

◎Making a chipset involves the use of many advanced technologies in circuit design and wafer fabrication, as evidenced by the dramatic increase in speed and performance of personal computers.

◎The global trend of reducing the amount of paper used in offices is spreading rapidly. For instance, many companies use their own Intranets and document management systems to effectively control the circulation of security documents.

◎An increasing number of interest rate derivative products, as priced by the forward interest rate, have highlighted the empirical challenges of fitting forward rate yield curves to current market data.

◎Consumer demand for hard disc bandwidth has significantly increased in recent years owing to the large size of files and increased disc size.

◎Magnesium alloys have great potential for diverse use in the automotive, railway, aerospace, computer, communications and consumer electronics industries since they have the lowest density of all metallic structural materials.

◎Taiwan's Ministry of Education is increasingly emphasizing the use of multi-assessment in middle schools and universities.

◎The benefits of web learning are widely recognized by educators and researchers.

◎Piecewise linearization algorithms are extensively used in nonlinear programming.

◎3D models are extensively adopted in multimedia applications owing to their relatively low cost and ease with which they can construct animated 3D objects.

◎Geographic-based information is increasingly used in daily lives.

◎Intelligent information retrieval systems utilize knowledge bases to increase the effectiveness of retrieval.

◎Geometric parameters affect the reliability of flip chip packaging under thermodynamic loading.

◎Many machine learning and optimization application-related problems are solved by GAs with the multi-state property.

**2. Work problem 工作問題. What is the problem that you are trying either to solve or more thoroughly to understand in the work proposal? 你的工作提案裡有你試著要解決或是想更進一步瞭解的問題嗎?**

Consider the following examples.

◎However, conventional methods that only focus on fitting a yield curve and transforming it into a forward rate curve result in an unreasonable, extremely high or negative forward rate.

◎The wireload model utilized in synthesis provides inadequate information regarding the routed net because it only estimates the delays based on the fan-out number and possible routing path.

◎Encouraging employees to view materials on-line instead of on paper is difficult.

◎Conventional parallel bus architecture has a relatively low transmission speed owing to interference between bus lines.

◎Exhaust in the form of visible, white smoke and an odor is occasionally emitted from factories . Identifying the origin of the white smoke and odor can be extremely complicated. For instance, the mixing reaction or coagulation of pollutants can form a small nucleus that reflects light and becomes invisible upon inspection. Additionally, some mixing reactions can cause an odor even though they are at the ppt level.

◎Conventional evaluation models are concerned only with economic factors and neglect factors that can not be evaluated in money.

◎Increasing demand on the reservoir supply in southern Taiwan will soon surpass the reservoir's capacity.

◎Scored methods inaccurately estimate the dispersion effect, confounding the location and dispersion effects.

◎Conventional mathematical models have difficulty in discriminating among multiple ranks.

◎Conventional control charts are designed for manufacturing systems with only one source of variation, making it impossible to control several independent sources of variation.

◎Uncertain and insufficient hydrology data make it impossible to predict accurately the amount of residuals in groundwater contaminants.

◎Conventional process capability indices (PCIs) can neither objectively assess quality and delivery time nor identify the relationship between PCIs and yield rate.

◎Magnesium alloys have a low formability near room temperature and are more expensive than plastics, aluminum, steel and cast iron.

**3. Quantitative specification of problem 問題的量化. How can you quantitatively define the problem so that the reader understands the quantitative limitations of previous efforts to solve this problem? 你要如何量化問題來讓你的讀者明白之前文獻研究所遇到的量化限制？**

Consider the following examples.

◎Testing time accounts for 20% of the total research and development time,

◎Doubling the test time decreases product fault coverage by only 5%.

◎While a loop filter with a wide bandwidth causes large vibration and ultimately a high error rate, a loop filter with a narrow bandwidth leads to slow which takes over ten times longer than the estimated acquisition time.

◎The conventional interface requires 28 pins and can only support 133 MHz bandwidths. Consequently, the failure rate in manufacturing exceeds 5% when the conventional interface must support 150 MHz bandwidths.

◎Testing in an emulator takes an extremely long time, for example, a benchmark taking 20 minutes in a real system requires 24 hours in an emulated system .

◎Piecewise linearization algorithms that require more than 10 hours to obtain the optimal solution for a general nonlinear programming problem make equipment-related costs too high.

◎A prediction accuracy of only 60% among simulated models exposes traders to unnecessarily high risks.

◎Conventional models require seven steps to construct a 3D model and their error rate exceeds 5%, making it impossible to widely commercialize animation software and construct realistic digital objects.

◎An error rate that exceeds 20% does not meet the R.O.C.'s environmental protection standards.

◎A situation in which the formability near room temperature does not reach 20% elongation limits the commercial application of magnesium and its alloys.

◎If the geometric parameters can not reach an accuracy of 5%, engineers can neither predict the fatigue life nor enhance the yield of a flip chip package.

◎Students who can implement only one GA in two weeks will learn GAs less effectively than those who can implement more.

◎The lack of uniform data for various dimensions of an order parameter makes it impossible to obtain the number of marginal dimensions that distinguishes the above scenarios.

◎The failure to understand precisely which microscopic features form a particular phase topology makes it impossible to analyze from first principles the behavior of simple model thermodynamic functions near the critical point.

◎Increasing the evaporation speed by 1% increases the resistance by 5%.

**4. Importance of problem 問題的中心. How will leaving the problem unsolved or not thoroughly understood negatively impact the reader of your work proposal? 如果問題沒被解決或是充分瞭解，這對工作提案的讀者會有多大的負面衝擊?**

Consider the following examples.

◎A reduction in power quality causes the DDR_SDRAM to perform unsatisfactorily. Moreover, resistor impedance and cross-talk cause noise in the PCB, negatively impacting the VGA Card's speed during operation and falling below FCC standards with respect to EMI.

◎An unstable or inaccurate measurement system will yield errors in process control and judgment, making it inappropriate in analyzing a manufacturing process.

◎A loop filter with a narrow bandwidth may not recover the carrier while the receiver suffers from a large frequency or phase offset, leading to failure in the digital receiver.

◎The time required for the verification process could increase exponentially as IC designs become more complex.

◎This high failure rate not only prevents the integration of many functions into a chip with limited pin counts because the conventional interface requires many pins, but also makes commercialization impossible.

◎Obtaining the number of marginal dimensions that distinguishes the above scenarios is impossible.

◎Analyzing from first principles the behavior of simple model thermodynamic functions near the critical point is impossible.

◎Estimating the radiation defect rebuilding in cesium halides doped with OH-impurities and measuring the ionic conductivity of irradiated crystals, are impossible.

◎The inability to understand fully their physical nature further delays experimental investigation and possible industrial application of these crystals.

◎The inability to understand fully the exoemission mechanism makes it impossible to detect and investigate defects in the surface layers of the crystals.

◎The resistance of thin films can not be prevented from fluctuating with a varying speed of evaporation.

◎The inability to understand fully the RSOE mechanisms in II-VI semiconductors makes it impossible to accurately predict the irradiation conditions and thus increase structural efficiency.

◎Use of the IMD Competitiveness Simulations may cause policy makers to make an inaccurate decision.

◎The inability of cooperation and competition to complement each other in an Internet-based learning environment makes it impossible to ensure that students participate equally.

◎Taiwan's educational system cannot multi-assess the higher-order thinking of students.

◎The inability of web collaborative design activity to promote students' higher order thinking makes it impractical to implement such an innovative means of instruction.

◎Wide commercialization of animation software and the construction of realistic digital objects are impossible.

**5. Project need 計畫需求. Based on the problem defined above, what is the immediate project need? 根據以上問題，最迫切的計畫需求是什麼? Consider the following examples.**

◎Therefore, an IC product with an excellent electrical / thermal performance, low cost, and high degree of reliability must be designed.

◎Therefore, an effective strategy must be developed in which our company's technical documents are saved into E-file format, and Notes Client is used to control all files.

◎Therefore, a novel model must be developed to derive directly a reasonable forward rate curve in one explicit function to correlate well with market data.

◎Therefore, an analysis method must be developed that (regardless of the complexity of the model(s) employed) provides accurate capacity data to the Production Planning Department to optimally plan wafer production, and to the Marketing Department to fully realize the fab's constraints.

◎Therefore, a serial interface must be developed, capable of easily supporting 150 MHz or higher bandwidth. Additionally, the pin counts of the interface must be decreased to ensure that a chip with many functions can be integrated.

◎Therefore, the global optimum must be obtained in general nonlinear programming models within a tolerable error and significantly increase computational efficiency by decreasing the use of 0-1 variables.

◎Therefore, a novel response surface methodology with simplified calculations must be developed to accurately estimate the location and dispersion effects, leading to an optimal combination of process parameters.

◎Therefore, the tolerable errors associated with constructing a digital face must be determined, enhancing multimedia or animation applications by reducing formulation costs and creating more realistic digital objects.

◎Therefore, the complexity of operation must be reduced, enabling an inexperienced engineer to implement a network management system without intensive training.

◎Therefore, a GA model must be developed, capable of enhancing the ability of conventional GAs to handle applications with a multi-state property.

◎Therefore, such systems with more than two equilibrium positions on a site must be accurately described, extending conventional order-disorder models to a wider class of materials.

◎Therefore, a novel response surface methodology with simplified calculations must be developed to estimate accurately the location and dispersion effects, leading to an optimal combination of process parameters.

**F** In the space provided, write a problem statement for your work proposal.

Setting of work proposal 工作提案建構:

_____

_____

_____

Work problem 工作問題:

_____

_____

Quantitative specification of problem 問題的量化:

_____

_____

Importance of problem  問題的中心:

_____

_____

Project need  計畫需求:

_____

_____

## G Look at the following examples of a problem statement for a work proposal.

*Industrial Engineering* 工業工程相關分類

**Setting of work proposal** 工作提案建構 Enterprises trade with each other by carefully considering business ratings to reduce investment risks. **Work problem** 工作問題 However, conventional mathematical models have difficulty in discriminating among multiple ranks. **Quantitative specification of problem** 問題的量化 A prediction accuracy of only 60% among past models **Importance of problem** 問題的中心 exposes traders to unnecessarily high risks. **Project need** 計畫需求 Therefore, a flexible and accurate neural network structure must be developed that applies artificial intelligence and fuzzy theory to business ratings and bankruptcy prediction.

*Civil Engineering* 土木相關分類

**Setting of work proposal** 工作提案建構 Demand on the water supply in Taiwan has skyrocketed owing to the island's increasing population and elevated living standards. **Work problem** 工作問題 Eventually, the current reservoir supply will be insufficient if current personal and industrial demand persists. **Importance of problem** 問題的中心 Consequently, this surging demand will lead to the increased drilling of new wells to locate groundwater resources. Such drillingis extremely expensive. **Project need** 計畫需求 An optimal operating model must therefore be developed, capable of choosing an effective monitoring well to adequately control a groundwater system.

*Industrial Engineering* 工業工程相關分類

**Setting of work proposal** 工作提案建構 Engineers heavily emphasize applicability and accuracy when using a process capability index to evaluate how the performance of a process. **Work problem** 工作問題 However, using conventional process capability indices to evaluate a non-normal distribution process often leads to inaccurate results. Additionally, calculating a point estimator with sampling data leads to inaccurately estimating the process capability index of the process population. Although sampling more data can increase the accuracy of an estimator, such an approach is costly and time consuming. **Importance of problem** 問題的中心 The above limitations cause engineers to make errors when comparing manufacturing processes or selecting an alternative supplier. **Project need** 計畫需求Therefore, a process capability index must be developed that adopts Clement's and the Bootstrap methods to overcome these obstacles.

*Civil Engineering* 土木相關分類

**Setting of work proposal** 工作提案建構 Although quantifying residual non-aqueous phase liquid (NAPL) contamination in the subsurface has received considerable attention, **Work problem** 工作問題 conventional methods such as core sampling and geophysical logging near contaminated sources can neither acquire sufficient samples nor identify where contaminants are distributed. **Quantitative specification of problem** 問題的量化 Consequently, the lack of effective predictive tools, an objective testing procedure and simulated models lead to inefficiency and high overhead errors. **Importance of problem** 問題的中心 Furthermore, improper use of those methods will yield inadequate/unsatisfactory/insufficient results and severe losses of time and cost money in industry. **Project need** 計畫需求 Therefore, an efficient hypothesis testing procedure for NAPLs must be developed, capable of assessing the amounts of residual NAPL contamination.

*Civil Engineering* 土木相關分類

**Setting of work proposal** 工作提案建構 Consumer demand on the water supply in southern Taiwan has significantly increased in recent years owing to population and industrial growth, and increased living standards. **Work problem** 工作問題 Although the reservoir supply in southern Taiwan can satisfy the current water demand, increasing demand will soon surpass the system's capacity. **Importance of problem** 問題的中心 Moreover, the lack of an effective water distribution strategy for current reservoirs will lead to frequent water shortages in southern Taiwan. **Project need** 計畫需求 An optimization model must therefore be developed, capable of effectively managing water resources for current reservoirs to decrease the probability of water shortages.

*Industrial Engineering* 工業工程相關分類

**Setting of work proposal** 工作提案建構 An increasing number of wafer fabs use a control chart to detect assignable causes, making it extremely difficult for engineers to control effectively the wafer process. **Work problem** 工作問題 However, conventional control charts are designed for manufacturing systems with only one source of variation, making it impossible to control several independent sources of variation. **Importance of problem** 問題的中心 Using such charts often leads to ineffective and futile searches for assignable causes, wasting considerable amounts of manpower and capital. **Project need** 計畫需求 Therefore, an efficient quality control process must be developed, capable of detecting assignable causes concealed behind both multiple characteristics and multiple readings in a manufacturing system with several sources of variation.

*Industrial Engineering* 工業工程相關分類

**Setting of work proposal** 工作提案建構 Most organizations evaluate business ratings subjectively, normally based on professional knowledge and experience. Most enterprises also lack simulation models capable of flexibly assessing trade partners' credit. **Work problem** 工作問題 However, neural networks accurately discriminate among multiple ranks, but some samples cannot be easily separated in ambiguous situations. **Quantitative specification of problem** 問題的量化 For instance, while applying Back Propagation Network and Fuzzy Theory to two ranks of investment and bankruptcy, Jiang (1998) forecasted business ratings using data from the previous three years, yielding accuracies of 85%, 75%, and 70%. **Importance of problem** 問題的中心 This low accuracy exposes traders to unnecessarily high risks. **Project need** 計畫需求 Therefore, the multiple ranks must be distinguished to achieve a forecasting accuracy of 90% for applications.

*Civil Engineering* 土木相關分類

**Setting of work proposal** 工作提案建構 Predicting the amount of residuals of ground water contaminants has been extensively studied. **Work problem** 工作問題 However, the amount cannot be predicted accurately **Quantitative specification of problem** 問題的量化 when errors exceed 20%, as caused by uncertain and insufficient hydrology data. **Importance of problem** 問題的中心 An error rate that exceeds 20% does not meet the R.O.C.'s environmental protection standards. **Project need** 計畫需求 Therefore, an NAPL simulator model must be developed that includes several parameters acquired by experimental data.

*Materials Science* 材料相關分類

**Setting of work proposal** 工作提案建構 Magnesium alloys have great potential for diverse use in the automotive, railway, aerospace, computer, communications and consumer electronics industries, since they have the lowest density of all metallic structural materials **Work problem** 工作問題 However, these alloys have a low formability near room temperature and are more expensive than plastics, aluminum, steel and cast iron. **Quantitative specification of problem** 問題的量化 A situation in which the formability near room temperature does not reach 20% elongation **Importance of problem** 問題的中心 limits the commercial application of magnesium and its alloys. **Project need** 計畫需求 A processing method must therefore be developed, capable of producing materials of a required shape at a relatively low cost, directly from wrought products.

*Statistics* 統計相關分類

**Setting of work proposal** 工作提案建構 Decision-making environments are increasingly complex, **Work problem** 工作問題 accounting for why conventional evaluation models are concerned only with economic factors and neglect those factors that can not be evaluated in money. **Quantitative specification of problem** 問題的量化 In addition to an economic criterion, for example, when evaluating different brands of bus systems, many criteria are still applicable, such as the level of pollution from the bus and the drivability of the bus. **Importance of problem** 問題的中心 Consequently, such models may yield an incomplete evaluation and decision makers may ultimately select an inappropriate scheme.  Moreover, implementing an inappropriate scheme will have large social costs and waste resources. **Project need** 計畫需求 Therefore, an appropriate evaluation model capable of selecting natural gas bus brands must be developed.

*Statistics* 統計相關分類

**Setting of work proposal** 工作提案建構 Despite their extensive use in solving problems related to ordered categorical data quality, **Work problem** 工作問題 conventional scored methods are unnecessarily complex and inaccurate in estimating the dispersion effect. **Quantitative specification of problem** 問題的量化 Additionally, their implementation leads to an inaccurate optimal combination of process parameters, **Importance of problem** 問題的中心 requiring much time and a higher cost in the product design stage. **Project need** 計畫需求 Therefore, a novel response surface methodology with simplified calculations must be developed to accurately estimate the location and dispersion effects, leading to an optimal combination of process parameters.

*Industrial Engineering* 工業工程相關分類

**Setting of work proposal** 工作提案建構 Although process quality and delivery time have been increasingly emphasized by industry, **Work problem** 工作問題 conventional process capability indices (PCIs) can neither objectively assess quality and delivery time nor identify the relationship between PCIs and yield rate. **Quantitative specification of**

**problem** 問題的量化 Consequently, the lack of an effective performance index and an objective procedure will lead to inefficiency and a high overhead cost. **Importance of problem** 問題的中心 Furthermore, firms that perform poorly in terms of quality and delivery will lose their market competitiveness. **Project need** 計畫需求 Therefore, an efficient hypothesis testing procedure for PCIS must be developed, capable of assessing the operational cycle time (OCT) and delivery time (DT) for VLSI.

*Distance Learning* 遠距教學相關分類

**Setting of work proposal** 工作提案建構 Although an increasing number of Internet-based learning environments are available to enhance knowledge construction, **Work problem** 工作問題 the learning activities of these learning environments overrely on in-group cooperation, erroneously implying that group members participate equally. **Importance of problem** 問題的中心 Additionally, the inability to allow cooperation and competition to complement each other in an Internet-based learning environment makes it impossible to ensure that students participate equally. **Project need** 計畫需求 Therefore, an Internet-based constructive learning environment must be developed to allow participants interactively to link their conceptual maps for accumulative learning.

*Computer Science* 資訊科學相關分類

**Setting of work proposal** 工作提案建構 Although an increasing number of mathematical courses are available on the Web, **Work problem** 工作問題 conventional editors using plain text as their user interface, have difficulty in editing complex mathematical equations. **Quantitative specification of problem** 問題的量化 Consequently, users of conventional editors spend an excess of time to express equations that include more than ten mathematical symbols, **Importance of problem** 問題的中心 causing students to spend longer to learn how to use the teaching system. **Project need** 計畫需求 Therefore, a user-friendly editor of mathematical symbols capable of using a graphic user interface, must be developed to reduce the learning time and provide an easier way to edit complex mathematical equations on the web.

*Distance Learning* 遠距教學相關分類

**Setting of work proposal** 工作提案建構 Despite their increasing availability on the Internet, **Work problem** 工作問題 distance learning courses lack feasible strategies for assessing student performance, ultimately inhibiting distance learning. Despite their limitations, some web-based testing systems have been implemented to evaluate students' learning outcomes. **Quantitative specification of problem** 問題的量化 For example, a student can answer questions on a test by studying prepared materials from a remote location. Notably, the drawbacks of a web-based testing system can be eliminated if a student's learning ability is evaluated through an activity. **Project need** 計畫需求 Therefore, a networked peer assessment system that can support instruction and learning must be developed to analyze students' learning outcomes in higher education.

*Computer Science* 資訊科學相關分類

**Setting of work proposal** 工作提案建構 Taiwan's Ministry of Education is increasingly emphasizing the use of multi-assessment in middle schools and universities. **Work problem** 工作問題 However, conventional methods of assessing students' abilities fail to assess higher-order thinking owing to their inability to motivate students properly. **Importance of problem** 問題的中心 Consequently, Taiwan's educational system can not multi-assess the higher-order thinking of students. **Project need** 計畫需求 Therefore, a networked electronic portfolio system with peer assessment must be developed to provide a creative means of assessing the higher-order thinking of students.

*Computer Science* 資訊科學相關分類

**Setting of work proposal** 工作提案建構 Although extensively used in nonlinear programming, **Work problem** 工作問題 piecewise linearization algorithms require too much time to obtain an optimum solution. For instance, while spending much time in attempting to solve DNA-related problems, biologists only obtain the local optimum in most cases. **Quantitative specification of problem** 問題的量化 If piecewise linearization algorithms require more than ten hours to obtain the optimal solution for general nonlienar programming problems, **Importance of problem** 問題的中心 then equipment-related costs involved in obtaining the optimal solution are too high. **Project need** 計畫需求 Therefore, the global optimum must be obtained in general nonlinear programming models to within a tolerable error. In doing so, computational efficiency can be significantly increased by decreasing the use of 0-1 variables.

*Civil Engineering* 土木相關分類

**Setting of work proposal** 工作提案建構 Groundwater usage in Taiwan is increasing at an accelerated rate, **Work problem** 工作問題 leading to a growing incidence of groundwater pollution in major groundwater supply regions in Taiwan, such as the Ping-tung Plain. Contaminants that pollute the aquifer make drinking the water from that groundwater source impossible for several years. **Project need** 計畫需求 Therefore, a deterministic and stochastic model must be developed for simulating groundwater flow to assess monitoring network alternatives.

*Information Management* 資訊管理相關分類

**Setting of work proposal** 工作提案建構 Although 3D models are extensively adopted in multimedia applications owing to their relatively low cost and ease with which they can construct animated 3D objects, **Work problem** 工作問題 conventional 3D models are too time consuming and inaccurate when constructing digital objects since they manually retrieve 2D images. Moreover, their complex transformation procedures make them too costly and complicated for implementation. **Quantitative specification of problem** 問題的量化 A situation in which conventional models require seven steps to construct a 3D model with an error rate that exceeds 5% **Importance of problem** 問題的中心 makes it impossible to widely commercialize animation software and construct realistic digital objects. **Project need** 計畫需求 Therefore, the tolerable errors associated with constructing a digital face must be determined, enhancing multimedia or animation applications by reducing formulation costs and creating more realistic digital objects.

*Information Management* 資訊管理相關分類

**Setting of work proposal** 工作提案建構 Despite the increasing use of geographic-based information in daily lives, **Work problem** 工作問題 the passive mode of accessing information fails to transmit effectively geographic-based information to PDA users. Although unaware at the time, PDA users may need information such in case of an emergency. **Importance of problem** 問題的中心 The inability of PDA users to receive updated information in a timely manner will limit PDA use to within a narrow range. Such a limitation may discourage PDA use. **Project need** 計畫需求 Therefore, a GIS-based architecture that supports more interesting and useful functions than those of conventional architecture must be developed.

*Information Management* 資訊管理相關分類

**Setting of work proposal** 工作提案建構 Intranets are extensively used by enterprises to accelerate commercial activities. **Work problem** 工作問題 However, conventional network management systems are too expensive and complicated for implementation in an enterprise's Intranet. **Quantitative specification of problem** 問題的量化 If the conventional network management system costs more than 300,000 US Dollars to implement and requires additional machinery to operate, **Importance of problem** 問題的中心 enterprises cannot upgrade hardware and software. **Project need** 計畫需求 Therefore, the complexity of operation must be reduced, enabling an inexperienced engineer to implement a network management system without intensive training.

*Computer Science* 資訊科學相關分類

**Setting of work proposal** 工作提案建構 Although an increasing number of Genetic Algorithm (GA) courses are offered to solve optimization problems, **Work problem** 工作問題 Students spend much time in coding programs for exercises when learning GAs, making it impossible to implement many GAs in a relatively short time. For instance, a student must be able to implement GAs beforehand. **Quantitative specification of problem** 問題的量化 For instance, students who can implement only one GA in two weeks **Importance of problem** 問題的中心 will learn GAs less effectively than those who can implement more. **Project need** 計畫需求 Therefore, the need for hand coding GA programs must be eliminated, simplifying the process of learning genetic algorithms.

*Computer Science* 資訊科學相關分類

**Setting of work proposal** 工作提案建構 Although intelligent information retrieval systems utilize knowledge bases to increase the effectiveness of retrieval, **Work problem** 工作問題 However, most knowledge bases are constructed by inquiring of domain experts to acquire knowledge. **Quantitative specification of problem** 問題的量化 For instance, a situation in which an average of one hour is required to construct a knowledge base that contains 50 concepts **Importance of problem** 問題的中心 results in poor efficiency. **Project need** 計畫需求 An automatic knowledge-base construction methodology must therefore be developed to improve on the conventional approaches.

*Computer Science* 資訊科學相關分類

**Setting of work proposal** 工作提案建構 Many machine learning and optimization application-related problems are solved by GAs with the multi-state property. For instance, in chess, a good player often employs various strategies based on his opponent's moves, the game's progress, or the chess clock. Therefore, an intelligent chess playing program should consider the multi-state property to perform more effectively. **Work**

**problem** 工作問題 However, conventional methods cannot solve multi-state problems. **Quantitative specification of problem** 問題的量化 If the solution varies with the problem state in a multi-state problem, **Importance of problem** 問題的中心 conventional methods neglect the multi-state property and thus yield an inaccurate and unfeasible solution. **Project need** 計畫需求 Therefore, a GA model must be developed, capable of enhancing conventional GAs in handling applications with a multi-state property.

*Mechanical Engineering* 機械相關分類

**Setting of work proposal** 工作提案建構 An increasing number of digital modules are embedded in high performance systems. **Work problem** 工作問題 However, the process, voltage, temperature, loading (PVTL) factors inevitably induce the clock-skew problem. Moreover, the skew problem will worsen as the clock's operational frequency increases, **Importance of problem** 問題的中心 becoming a bottleneck in future-high-performance systems and possibly resulting in a system's malfunctioning. **Project need** 計畫需求 Therefore, a SAR-controlled DLL deskew circuit must be developed to alleviate the system clock skew problem.

*Mechanical Engineering* 機械相關分類

**Setting of work proposal** 工作提案建構 Thermodynamic loading affects the reliability of flip chip packaging under thermodynamic loading, **Work problem** 工作問題 making it impossible to predict accurately the geometric parameters of a C4 type solder joint. **Quantitative specification of problem** 問題的量化 The inability of geometric parameters to reach an accuracy of 5% **Importance of problem** 問題的中心 makes it impossible to predict the fatigue life and enhance the yield of a flip chip package. Therefore, **Project need** 計畫需求 geometric parameters of a C4 type solder joint must be predicted more accurately than by conventional methods.

*Physics* 物理相關分類

**Setting of work proposal** 工作提案建構 As well known, weak disorder leads to two critical behavior scenarios of the O(m) model. **Work problem** 工作問題 However the critical behavior of materials described by the weakly diluted O(m) model remains unclear with respect to the dimensions of the order parameter. **Importance of problem** 問題的中心 For various dimensions of an order parameter on which no uniform data is available, obtaining the number of marginal dimensions that distinguish the above scenarios is impossible. **Project need** 計畫需求 Therefore, a procedure for estimating mc must be proposed using well known approaches and must have a lower error bar on the value of mc than other approaches found in the literature.

*Physics* 物理相關分類

**Setting of work proposal** 工作提案建構 Dielectric properties of

DMAGaS-DMAAlS crystals produce variations in temperature and pressure behavior. **Work problem** 工作問題 Such behavior accounts for the lack of clarity on the physical nature of dielectric properties of the ferroelectrics. These crystals have a similar structure, but different phase diagrams. The antiferroelectric phase is observed only in the DMAGaS crystals. **Importance of problem** 問題的中心 If the physical nature of the dielectric properties of both ferroelectrics remains unclear, further experimental investigation and possible industrial application of these crystals will be delayed. **Project need** 計畫需求 Therefore, such systems with more than two equilibrium positions on a site must be accurately described, extending conventional order-disorder models to a wider class of materials.

*Physics* 物理相關分類

**Setting of work proposal** 工作提案建構 The surface control method, based on exoelectron emission (EEE), is a precise and nondestructive relaxational method and can detect and predict the early stages of the destruction of materials. Additionally, exoemission from alkali halide crystals has been extensively studied theoretically and experimentally. **Work problem** 工作問題 However, the theoretical description of exoemission from alkali halide crystals is insufficient, making it difficult to analyze and interpret exoemission current-related data. **Importance of problem** 問題的中心 The inability to understand fully the exoemission mechanism makes it impossible to detect and investigate defects in the surface layers of the crystals. **Project need** 計畫需求 Therefore, exoelectronic energy spectra should be theoretically calculated to determine whether the defect-recombination mechanism is responsible for EEE from CsBr.

*Physics* 物理相關分類

**Setting of work proposal** 工作提案建構 As is well known, radiation induces defect rebuilding. **Work problem** 工作問題 However, the exact process by which radiation impacts ionic conductivity remains unknown. **Importance of problem** 問題的中心 The inability to understand its impact on ionic conductivity makes it impossible not only to estimate the radiation defect rebuilding in cesium halides doped with OH-impurities, but also to measure the ionic conductivity of irradiated crystals. **Project need** 計畫需求 Therefore, the formation of radiation-stimulated defects in doped cesium iodide crystals must be investigated.

*Physics* 物理相關分類

**Setting of work proposal** 工作提案建構 The interaction between binary mixture components leads to diverse phase behavior with respect to relative molecular sizes and the strengths of their interactions. **Work problem** 工作問題 However, fully understanding the relationship between the microscopic description and macroscopic phase behavior in a binary liquid system is extremely difficult. **Quantitative specification of problem** 問題的量化 Additionally, the failure to understand precisely which microscopic features form a particular phase topology **Importance of problem** 問題的中心 makes it impossible to analyze from first principles the behavior of simple model thermodynamic functions near the critical point. **Project need** 計畫需求 Therefore, the critical behavior of a binary symmetrical mixture must be elucidated using the collective variables method.

*Physics* 物理相關分類

**Setting of work proposal** 工作提案建構 Evaporation methods enhance the formation of scattering centers and the electrical conductivity of thin metal films. **Work problem** 工作問題 However, increasing the evaporation speed increases the resistance of metal films. **Quantitative specification of problem** 問題的量化 For instance, increasing the evaporaion speed by 1% increases the resistance by 5%, **Importance of problem** 問題的中心 making it impossible to prevent the resistance of thin films from fluctuating with a varying speed of evaporation. **Project need** 計畫需求 Therefore, the exact process by which scattering centers influence the electrical conductivity of metal films must be investigated using a novel evaporation method.

*Physics* 物理相關分類

**Setting of work proposal** 工作提案建構 As well known, ionizing

irradiation causes material damage. **Work problem** 工作問題 However, exactly why the RSOE mechanism is present in different materials remains unknown. **Importance of problem** 問題的中心 Additionally, the inability to understand the fully RSOE mechanisms in II-VI semiconductors makes it impossible to accurately predict the irradiation conditions and increase structural efficiency. **Project need** 計畫需求 Therefore, in addition to investigating the role of RSOE in II-VI semiconductors, a related model must be constructed.

*Physics* 物理相關分類

**Setting of work proposal** 工作提案建構 Dipole relaxation arises in doped CsJ crystals. **Work problem** 工作問題 However, whether radiation decay of I-V dipoles occurs in CsJ crystals remains unknown. **Importance of problem** 問題的中心 Additionally, the inability to understand fully the formation of radiation-stimulated defects in doped CsJ crystals makes it impossible to use these crystals in technological elements. **Project need** 計畫需求 Therefore, an attempt must be made to confirm defect-recombination mechanism as appropriate for exoemission from CsBr - similar structures

*Information Management* 資訊管理相關分類

**Setting of work proposal** 工作提案建構 Taiwan's global competitiveness ranking in the IMD World Competitiveness Scoreboard is falling. **Work**

**problem** 工作問題 The IMD Competitiveness Simulations neglect interactions among related factors. **Importance of problem** 問題的中心 Use of the IMD Competitiveness Simulations possibly causes policy makers to make inaccurate decisions. **Project need** 計畫需求 Therefore, a novel competitiveness model must be developed, capable of integrating a global optimization algorithm into the IMD World Competitiveness Model.

*Information Management* 資訊管理相關分類

**Setting of work proposal** 工作提案建構 The results of testing bugs in IC design are commonly submitted to a database. **Work problem** 工作問題 However, the dBase format often causes malfunctioning in the database after three years of use, **Quantitative specification of problem** 問題的量化 as evidenced by the increasing frequency of corruption of the database. **Importance of problem** 問題的中心 Losing statistical records while repairing the corrupted database would be devastating to the quality of our products. Additionally, the inability of the dBase format automatically to backup records is worrying. Although database information can be backed up manually in a timely manner, human error or oversight could lead to a loss of records. **Project need** 計畫需求 Therefore, the original simple database type must be transferred to SQL format, which can support a large number of records.

*Electrical Engineering* 電子相關分類

**Setting of work proposal** 工作提案建構 Product life cycles in the information industry are shortening. Increasing market share depends on timely delivering of quality products. **Work problem** 工作問題 However, good product quality requires considerable spent on testing, during research and development. **Quantitative specification of problem** 問題的量化 If testing time accounts for 20% of the total research and development time, **Importance of problem** 問題的中心 then R&D engineers can not focus on writing code, reducing product quality and delaying production. **Project need** 計畫需求 Therefore, an automotive testing system capable of overcoming such a shortage must be developed.

*Electrical Engineering* 電子相關分類

**Setting of work proposal** 工作提案建構 The quality of wafers is essential, especially in light of the shrinking of electronic devices and the complexity of circuits in a chipset. The test period is prolonged to ensure that most functional paths are detected before mass production. **Work problem** 工作問題 However, the rising cost of tests has not improved overall testing. **Quantitative specification of problem** 問題的量化 For example, doubling the test time decreases product fault coverage by only 5%. **Importance of problem** 問題的中心 The lack of major test parameters and methods makes it impossible to enhance testing. **Project need** 計畫需求 Therefore, effective test parameters must be developed and their effectiveness demonstrated.

### *Electrical Engineering* 電子相關分類

**Setting of work proposal** 工作提案建構 High speed is an increasingly important feature of the 3D VGA Card in the PC industry. **Work problem** 工作問題 However, in a high speed PC, the resistor impedance, cross_ talk , power quality, and EMI are problematic for further development. **Quantitative specification of problem** 問題的量化 The inability to resolve these problems lowers the speed and overall performance of PCs. **Importance of problem** 問題的中心 Additionally, a reduction in power quality causes the DDR_SDRAM to perform unsatisfactorily. Moreover, resistor impedance and cross- talk cause noise in the PCB, negatively impacting the VGA Card's speed during operation and falling below FCC standards with respect to EMI. **Project need** 計畫需求 Therefore, a better quality 3D VGA board with less noise must be developed that is capable of matching the resistor impedance as well as reducing the EMI in the VGA board.

### *Industrial Engineering* 工業工程相關分類

**Setting of work proposal** 工作提案建構 Engineers often decide whether to adjust a manufacturing process according to measurement data. The quality of measurement data is related to the measurement system and its environment. **Work problem** 工作問題 However, conventional methods fail to establish the stability of a measurement system; specify how to repeat, reproduce and match a measurement system, or provide a criterion of judgment. **Importance of problem** 問題的中心 An unstable or inaccurate measurement system will yield  errors in process control and judgment, making it inappropriate in analyzing a manufacturing process. **Project need** 計畫需求 Therefore, an accurate measurement method capable of assessing the capability of a measurement system must be developed.

*Computer Graphics* 電腦製圖相關分類

**Setting of work proposal** 工作提案建構 Stereo 3D displays play an increasingly prominent role in computer graphics, visualization, and virtual-reality systems. **Work problem** 工作問題 However, conventional methods inefficiently derive stereoscopic images, owing to the complex geometrical computation required. **Importance of problem** 問題的中心 The inability to resolve this inefficiency problem makes high-quality and real-time 3D applications impossible. **Project need** 計畫需求 Therefore, a method of converting a 3D display to a stereo 3D display, using compatible hardware and software without special hardware.

*Electrical Engineering* 電子相關分類

**Setting of work proposal** 工作提案建構 The current trend of decreasing the length of a metal line in advanced IC processes has made the gap fill ability of HDP-CVD increasingly important in IC backend processes. **Work problem** 工作問題 However, the conventional HDP-CVD process is characterized by a high Ar gas flow and pressure. **Quantitative specification of problem** 問題的量化 A situation in which Ar gas flow and pressure exceed acceptable levels **Importance of problem** 問題的中心 easily causes re-deposition in the high aspect ratio structure and decreases the gap fill ability. **Project need** 計畫需求 An HDP-CVD process must therefore be developed, capable of reducing the Ar gas flow from 390 sccm to 50 sccm and the pressure from 5 mtorr to 2.5 mtorr.

*Electrical Engineering* 電子相關分類

**Setting of work proposal** 工作提案建構 Despite a large frequency offset, wide range locking with fast acquisition circuit-designed carrier recovery, helps a digital receiver to lock the carrier frequency in a short time with a tolerable error rate. **Work problem** 工作問題 However, conventional methods cannot do so just by utilizing digital a phase-locked loop (PLL) circuit  since the loop filter is a one-order low-pass filter. **Quantitative specification of problem** 問題的量化 While a loop filter with a wide bandwidth causes large vibration and ultimately a high error rate, a loop filter with a narrow bandwidth leads to slow convergence that takes over ten times longer than the estimated acquisition time. **Importance of problem** 問題的中心 Furthermore, a loop filter with a narrow bandwidth may not recover the carrier while the receiver suffers from a large frequency or phase offset, leading to failure in the digital receiver. **Project need** 計畫需求 Therefore, a method with additional apparatus must be designed to select efficiently the bandwidth of the loop filter.

*Electrical Engineering* 電子相關分類

**Setting of work proposal** 工作提案建構 Wafer product life is becoming gradually shorter as the conductor line width in integrated circuits decreases from the sub-micro to deep sub-micro level. **Work problem** 工作問題 As the reliability and lifetime of wafer products continue to fall, **Importance of problem** 問題的中心 production falls below customers' specifications. **Project need** 計畫需求 Therefore, an IC aging test system must be developed in which aging and burning tests can identify wafer defects.

*Quality Control* 品保相關分類

**Setting of work proposal** 工作提案建構 Some IC designs have become commercially available without the entire verification process's being executed, reducing the time to market. **Work problem** 工作問題 However, this approach creates potentially large financial risks for an IC design firm. **Quantitative specification of problem** 問題的量化 For instance, in 1995, Intel spent nearly 500 million US dollars in recalling Pentium CPUs that contained one floating-point division bug. **Importance of problem** 問題的中心 Unfortunately, the time required for the verification process could

133

increase exponentially as IC designs become more complex. **Project need** 計畫需求 A verification process must therefore be developed, capable of ensuring the accuracy of an IC design in a relatively short time.

*Electrical Engineering* 電子相關分類

**Setting of work proposal** 工作提案建構 Electrical parameters significantly affect wafer quality. Additionally, the acceptance criterion for these parameters is A2/R3, that is, three of five points tested must be pass for a wafer to be accepted. **Work problem** 工作問題 Moreover, the inability to monitor online whether products are out of specification increases the likelihood of poor wafer quality, **Importance of problem** 問題的中心 endangering a company's competitive edge. **Project need** 計畫需求 An on-line SPC system must therefore be constructed, capable of detecting OOS (out of specification) points and displaying the message by e-mail to the owner on time.

*Computer Graphics* 電腦製圖相關分類

**Setting of work proposal** 工作提案建構 3D applications play an increasingly important role in daily life, and especially in entertainment. **Work problem** 工作問題 However, improvements in software and hardware can not keep pace with consumer preferences. **Importance of problem** 問題的中心 The inability to design a high performance graphic chip bounded with drivers severely limits 3D applications. **Project need** 計

畫需求 Therefore, a high quality 3D graphics driver must be developed with excellent performance.

*Electrical Engineering* 電子相關分類

**Setting of work proposal** 工作提案建構 Consumer demand for hard disc bandwidth has significantly increased in recent years owing to the large size of files, and increased disc size. **Work problem** 工作問題 However, the conventional interface for hard discs cannot easily support such a high bandwidth and demands many pin counts to support such a requirement. **Quantitative specification of problem** 問題的量化 The conventional interface requires 28 pins and can only support 133 MHz bandwidth. Consequently, the failure rate in manufacturing exceeds 5% when the conventional interface must support 150 MHz bandwidth. **Importance of problem** 問題的中心 prevents the integration of many functions into a chip with limited pin counts because the conventional interface requires many pins, but also makes commercialization impossible. **Project need** 計畫需求 Therefore, a serial interface must be developed, capable of easily supporting 150 MHz or higher bandwidth. Additionally, the pin counts of the interface must be decreased to ensure that a chip can be integrated with many functions.

*Environmental Engineering* 環工相關分類

**Setting of work proposal** 工作提案建構 Wafer fabrication involves many chemicals that pose a potential threat to humans and the environment. A wafer fab must accumulate the process exhaust and reduce the concentration of pollutants to an acceptable level and thus adhere to environmental protection laws. **Work problem** 工作問題 Nevertheless, exhaust in the form of visible, white smoke and an odor is occasionally emitted from factories. Identifying the origin of the white smoke and odor can be extremely complicated. For instance, the mixing reaction or coagulation of pollutants can form a small nucleus that reflects light and becomes invisible. Additionally, some mixing reactions can cause an odor even though they are at the ppt level. **Importance of problem** 問題的中心 The inability to reduce the visible, white smoke and odor to acceptable levels creates not only a potentially unsafe working environment but also poses a perceived threat to the nearby community. **Project need** 計畫需求 Therefore, a strategy must be developed to maintain the continuity of operation, and the problem must be both forecast and solved in advance.

*Industrial Engineering* 工業工程相關分類

**Setting of work proposal** 工作提案建構 Capacity planning is essential in evaluating a wafer fab's capacity. For instance, a wafer fab's capacity is forecasted six months ahead to measure its productivity. **Work problem** 工作問題 However, conventional analysis models can not accurately forecast capacity in the short (dynamic) and long (static) term owing to the varying complexity of the model(s) employed. **Importance of problem** 問題的中

心 The inability to forecast accurately capacity in the short (dynamic) and long (static) term, regardless of the complexity of the model(s) employed, will ultimately lower a company's market competitiveness. **Project need** 計 畫需求 Therefore, an analytic method must be developed that (regardless of the complexity of the model(s) employed) provides accurate capacity data to the to optimally plan wafer production and to the Marketing Department in order to realize fully the fab's constraints through these capacity data.

*Electrical Engineering* 電子相關分類

**Setting of work proposal** 工作提案建構 The micron-level semiconductor process is characterized by a limited number of transistors and the occupied area. Compared with the intrinsic delay of cells, the net delay is insignificant enough to be overlooked. In contrast, the deep sub-micron design involves cells with an enhanced performance and slight delay. Therefore, net delay rather than cell delay is becoming the dominating factor in the semiconductor process. **Work Problem** 工作問題 Unfortunately, the wireload model utilized in synthesis provides inadequate information regarding the routed net because it only estimates the delays based on the fanout number and possible routing path. **Quantitative specification of problem** 問題的量化 For instance, the inaccuracy of wire delays based on wireload models ranges from roughly 10% to 30%. **Importance of problem** 問題的中心 Importantly, neglecting the assigned locations of the cells are physically assigned during synthesis makes it impossible to evaluate efficiently the wire delays and achieve optimization. **Project need** 計畫需求  An algorithm must therefore be developed, capable of omitting redundant and time-consuming steps when applying design automation for wafer chips.

*Electrical Engineering* 電子相關分類

**Setting of work proposal** 工作提案建構 Consumer demand for high speed transmission between processors and peripheral storage devices in computer systems has significantly increased in recent years. **Work problem** 工作問題 However, conventional parallel bus architecture has a relatively low transmission speed owing to interference between bus lines. **Quantitative specification of problem** 問題的量化 For instance, while the parallel bus hard-disk can only operate at 100 MB/sec, the serial one can operate between 150 MB/sec and 600 MB/sec. **Importance of problem** 問題的中心 Consequently, the lack of a high speed storage device decreases the efficiency of the entire computer system. **Project need** 計畫需求 Therefore, a serial hard-disk architecture must be developed for an actual wafer product, capable of operating at high speeds.

*Finance* 財務相關分類

**Setting of work proposal** 工作提案建構 An increasing number of interest rate derivative products priced by the forward interest rate have highlighted the empirical challenges of fitting forward rate yield curves to current market data. **Work problem** 工作問題 However, conventional methods, which only focus on fitting a yield curve and transforming it into a forward rate curve, result in an unreasonable, extremely high or negative forward rate, **Quantitative specification of problem** 問題的量化 such as 150% and -60%, respectively. **Importance of problem** 問題的中心 Pricing interest rate derivatives by adopting these inappropriate fitting forward rate methods will result in large deviations from market data. **Project need** 計畫需求 Therefore, a novel model must be developed to derive directly a reasonable forward rate curve in one explicit function to correlate well with market data.

*Quality Control* 品保相關分類

**Setting of work proposal** 工作提案建構 The global trend of reducing the amount of paper used in offices is spreading rapidly. For instance, many companies use their own Intranets and document management systems to control effectively the circulation of security documents. **Work problem** 工作問題 However, encouraging employees to view materials on-line instead of on paper form is difficult. **Importance of problem** 問題的中心 Additionally, paper usage involves too much time in acquiring necessary signatures for a particular document. **Project need** 計畫需求 An effective strategy must be developed in which our company's technical documents are saved in E-file format, and Notes Client is used to control all files.

*Electrical Engineering* 電子相關分類

**Setting of work proposal** 工作提案建構 High performance IC package design has become increasingly complex in current applications. **Work problem** 工作問題 However, the conventional notion of simply "packing" the IC can not satisfy current requirements. **Importance of problem** 問題的中心 For instance, IC package design that does not consider high performance, low cost and reliability will lead to a fall in an IC design company's competitiveness. **Project need** 計畫需求 Therefore, an IC product with an excellent electrical / thermal performance, low cost, and high degree of reliability must be designed.

*Electrical Engineering* 電子相關分類

**Setting of work proposal** 工作提案建構 Making a chipset involves the use of many advanced technologies in circuit design and wafer fabrication, as evidenced by the dramatic increase in the speed and performance of personal computers. **Work problem** 工作問題 However, the increasingly compact size of wafer chipsets causes problems in design and process. **Quantitative specification of problem** 問題的量化 Although increasing the wafer size and reducing the device's dimensions are strongly desired, a deep submicron process, 0.15um or lower, **Importance of problem** 問題的中心 will decrease not only the product yield and quality, but also the company's technical and market position. In determining the appropriate design and process, IC designers must adopt effective wafer design processes to increase product yield and lower overhead costs. **Project need** 計畫需求 Therefore, a strategy must be developed to solve the design and production-related problems of a chipset.

### *Electrical Engineering* 電子相關分類

**Setting of work proposal** 工作提案建構 As IC designs become increasingly complicated and larger, hardware emulation is essential for their verification. **Work problem** 工作問題 However, emulators such as our emulation solution are too expensive and too slow. **Quantitative specification of problem** 問題的量化 A Quickturn emulator costs more than a million US dollars and only works below 1 MHz, which is markedly lower than the frequency required in newly developed devices. **Importance of problem** 問題的中心 Such a high cost prevents us from verifying the designs of platforms, and the low speed makes testing difficult and the results unrealistic, because most devices, which are designed for 66MHz, 100MHz and 133MHz systems, cannot function normally in the 1MHz environment. Moreover, testing in an emulator takes an extremely long time such that a benchmark that requires 20 minutes in a real system requires 24 hours to complete in an emulated one. **Project need** 計畫需求 Therefore, an efficient emulation method must be developed that costs less and performs better than the conventional one.

*Electrical Engineerin* 電子相關分類

**Setting of work proposal** 工作提案建構 A reliable wafer chipset must not only have complementing functions, but also be efficient. In particular, the ability to forecast the life of a chipset is a priority concern in wafer product quality and reliability. **Work problem** 工作問題 However, conventional burn-in boards do not have complementing functions and are expensive. **Importance of problem** 問題的中心 The inability to forecast accurately the life cycle of a wafer chipset lowers a product's reliability. **Project need** 計畫需求 Therefore, a method must be developed to reduce the operating costs of the conventional burn-in board.

*Industrial Engineering* 工業工程相關分類

**Setting of work proposal** 工作提案建構 Despite the obstacles to accurately forecasting capacity planning, our company strives to enhance its market competitiveness by developing a more precise model. **Work problem** 工作問題 Although a viable solution to this problem, the dynamic capacity model requires a tremendous amount of data input. Generally, more data input implies a more accurate model. Restated, an accurate capacity forecast depends on sufficient input. According to our estimates, the dynamic capacity forecast and the actual throughput diverge by less than 10%. **Quantitative specification of problem** 問題的量化 Given the necessity of a stable server, a situation in which the server is down more than twice monthly will negatively impact our data reliability. **Importance of problem** 問題的中心 As well known, the CIM system is occasionally unstable. The deviation may require close collaboration between departments. **Project need** 計畫需求 Therefore, to solve this deviation problem, a precise dynamic capacity model capable of accurate forecasting capacity planning must be developed.

*Chemistry* 化學相關分類

**Setting of work proposal** 工作提案建構 Owing to environmental concerns, NF3 gas is gradually replacing CxFy gas in the dielectric CVD clean process. **Work problem** 工作問題 However, the global shortage of NF3 clean gas accounts for its costing significantly than CxFy clean gas, especially for high impurity ( > 4N ) NF3 clean gas. **Quantitative specification of problem** 問題的量化 Replacing CxFy clean gas costs more than three times more than using high impurity NF3 clean gas. Used in our fab's DCVD clean process for quite some time, high impurity NF3 clean gas is an expensive clean gas that creates a high overhead cost for our products.**Project need** 計畫需求 Therefore, a low impurity ( 3N ) NF3 clean gas must be developed, capable of reducing costs in the DCVD clean process.

# *Unit Four*

 Patrick  Information Management(資訊管理)

 Sally  Industrial Engineering(工業工程)

# Writing the hypothesis statement

假設描述

### Vocabulary and related expressions  相關字詞

formulate 使（表達影像）公式
3D image 3D影像
simultaneously 同時地
digitally construct 數位的製作
optimizing 最佳化
manually retrieving 手動獲得
tolerable 可容忍的
multimedia or animation applications 多媒體或卡通的運用
realistic digital objects 逼真的數位物體
evaluation model 評價模式
brands 廠牌
natural gas buses 天然氣巴士
cost effectiveness analysis 成本效能分析
criterion (判斷的)標準
multiple attribute decision making (MADM) 多目標規劃法
technique for order preference by similarity to ideal solution (TOPSIS) 逼近法

analytic hierarchy process 分析層級隨機過程
viable alternatives 可實行的其他方案
ranking methodology 排序法
objective outcome 客觀的成果
a more flexible procedure 一個更有彈性的程序
decision makers 決策者
order preferences
cesium iodine 碘化銫
cesium bromide crystals 溴化銫結晶體
doped with Oh impurity 摻添Oh移走電子
Stockbarger method in vacuum(真空) and Kyropolous method in air.
Gamma irradiation $\gamma$ 射線
conductivity 導電性
thermal stabilities 熱量的穩定
radiation defects 輻射衰敗

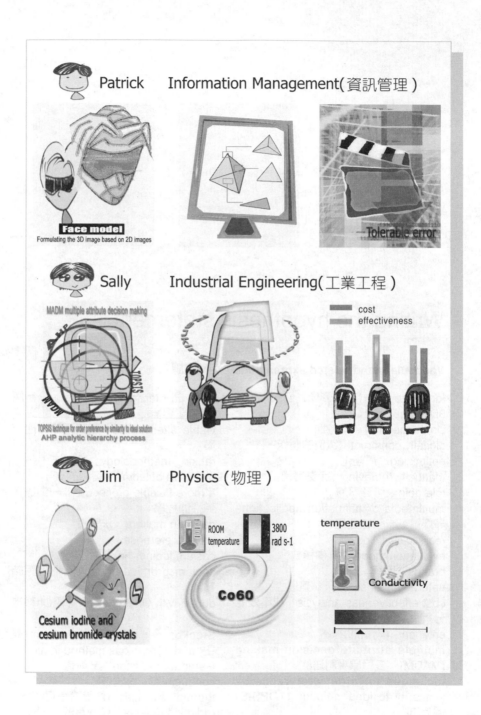

Patrick    Information Management(資訊管理)

**Face model**
Formulating the 3D image based on 2D images

Tolerable error

Sally    Industrial Engineering(工業工程)

MADM multiple attribute decision making

cost
effectiveness

TOPSIS technique for order preference by similarity to ideal solution
AHP analytic hierarchy process

Jim    Physics(物理)

Cesium iodine and
cesium bromide crystals

ROOM temperature    3800 rad s-1

Co60

temperature

Conductivity

**A** Write down the key points of the situations on the preceding page while the instructor reads aloud the script on page 332.

**Situation 1**

_____

_____

_____

**Situation 2**

_____

_____

_____

**Situation 3**

_____

_____

_____

**B** Based on the three situations in this unit, write three questions beginning with **Why**, and answer them.

**Examples**

*Why is Sally's proposed ranking methodology better than conventional ones?*

*It can provide a more objective outcome, including weights of related decision groups.*

*Why would decision makers prefer the ranking outcome of Sally's proposed methodology over that of other procedures?*

*Because it can identify the ordered preferences for alternatives*

1. _____

_____

2. _____

_____

3.

_____

_____

**C** Based on the three situations in this unit, write three questions beginning with ***What***, and answer them.

**Examples**

*What is Patrick's methodology for achieving his objective?*

*Three 2D images can be obtained simultaneously by using three standard cameras. The images can then be transferred to a personal computer. Additionally, the proposed model can be used to formulate digitally the 3D face.*

*What does Jim anticipate that the results of his study can provide further insight into?*

*The peculiarities of rebuilding radiation defects in cesium halides doped with OH impurities*

1.

_____

_____

2.

_____

_____

3.

_____

_____

**D** Based on the three situations in this unit, write three questions beginning with ***How***, and answer them.

**Examples**

*How does Sally plan to evaluate each criterion?*

*In terms of cost and effectiveness*

*How does Jim plan to measure the conductivity of both irradiated and non-irradiated crystals?*

*As a function of temperature*

1.

   _____

   _____

2.

   _____

   _____

3.

   _____

   _____

## E  Write questions that match the answers provided.

**Examples**

*What can Sally's proposed model evaluate?*

*The relationship between the cost and effectiveness of all viable alternatives to   bus systems*

*How does Patrick plan to obtain three 2D images simultaneously?*

*Using three standard cameras*

1.

_____

_____

Cesium iodine and cesium bromide crystals doped with OH impurities

2.

_____

_____

The peculiarities of rebuilding radiation defects in cesium halides doped with OH impurities

3.

_____

_____

The conductivity of both irradiated and non-irradiated crystals

# Unit Four

## Writing the hypothesis statement
## 假設描述

1 Work objective 工作目標
2 Methodology to achieve objective 達成目標的方法
3 Anticipated results 希望的結果
4 Contribution to field 領域的貢獻

Writing a hypothesis statement for the work proposal involves the following parts.

**1. Work objective 工作目標. What is the objective of your work proposal? 工作提案的目標？**

Consider the following examples.

◎An efficient response surface method can be developed, capable of optimizing ordered categorical data process parameters since setting the process parameters leads to optimization of the location and dispersion effects.

◎An efficient hypothesis testing procedure for PCIS can be developed, capable of assessing the operational cycle time (OCT) and delivery time (DT) for VLSI.

◎An Internet-based constructive learning environment can be developed that allows participants to link interactively their concept maps for accumulative learning.

◎A NAPL simulator model that includes several parameters acquired by experimental data can be developed.

◎A mathematical editor using the Java applet on the Web can be developed, capable of using a graphic user interface to edit mathematical symbols.

◎A networked electronic portfolio system with peer assessment can be developed to provide a creative means of assessing the higher-order thinking of students.

◎Online discussions of collaborative teams can be analyzed to reveal whether such activity engages students in higher order thinking and how it takes place.

◎An enhanced piecewise linearization algorithm capable of obtaining the global optimum of a nonlinear model can be developed for use in a web based optimization system.

◎An efficient face model capable of formulating the 3D image of an individual's face from three 2D images can be developed.

◎A novel web-based system in a PC-LAN environment capable of detecting network problems can be developed.

◎A novel learning environment can be developed, capable of assisting students in flexibly learning genetic algorithms based on computer-assisted instruction.

◎A deterministic and stochastic model for simulating groundwater flow can be developed to assess monitoring network alternatives.

◎A numerical groundwater model can be developed to analyze the position and the quantities of optimal monitor wells in Choshuichi River Fan.

◎An appropriate process capability index can be developed to evaluate non-normal distribution processes.

**2. Methodology to achieve objective 達成目標的方法　What are the steps by which your will achieve the above objective? 你的計畫中達成目標的步驟?**

Consider the following examples.

◎This novel methodology with simplified calculations of the mean and standard deviation of ordered categorical responses can be used to estimate the location and dispersion effects. Additionally, regression models can be used to relate the location and dispersion effects towards the controlled factor levels. Finally, an optimal level combination of process parameters can be obtained using the dual response surface method.

◎The PCIs of OCT and DT can be defined, and the UMVU (uniformly minimum variance unbiased) estimators of the studied PCIs can be derived under the assumption of a normal distribution. The above estimators can then be used to construct a one-to-one relationship between the PCIs and the conforming rate of DT (or OCT). Finally, a hypothesis testing procedure can be developed for the PCIs.

◎Three groups (each comprised of 10 or 11 students) can be formed. and a chapter assigned to each group for reading and concept mapping. Instructed to construct individual concept maps in a knowledge construction environment, each student can only view his or her own concept map to prevent students from copying their peers. Next, after reading concept maps constructed by the other groups, students can select the best concept map. Students can be instructed to link their concept maps to the best one. The concept maps of each student can then be evaluated by voting.

◎A web-based optimization system can be implemented based on the enhanced algorithm, using a dynamic linking library procedure. The system can then be linked to many other mathematical methods, for example, LINGO, to solve a nonlinear problem by integrating concurrent methods. Next, user specified problems can be  stored in a database storage system. Additionally, the solution can be derived to guarantee the global optimum with an acceptable error rate.

◎A graphic interface can be designed to illustrate the network structure of an enterprise. A free software program known as a Multi Router Traffic Grapher (MRTG) can then be applied to communicate with network devices. Return values from network devices can then be filtered to detect potential network problems. Additionally, the visualized information can be updated on the user interface of a web page.

◎The differential equation of curvature can be derived from the well-known Laplace-Young equation. Additionally, the contact angle between the solder and substrate and the surface tension of the solder can be measured. The free surface energy of the substrate and the interfacial tension between the solder and the substrate can then be obtained from Young's equation. Next, the differential equation of curvature is numerically solved to obtain the geometric parameters. The geometric data of the solder joint can be further used as an input in the finite element model to analyze the stress/strain distribution, thermal fatigue life and reliability of the electronic packages.

◎The grand partition function of a binary symmetrical mixture can be constructed near the vapour-liquid and mixing-demixing critical point. The grand partition function can then be reduced to the partition function of a three-dimensional Ising model in an external field. Next, for a hard-sphere square-well symmetrical binary mixture, the thermodynamic characteristics can be calculated as a function of microscopic parameters.

◎The model Hamiltonian can be constructed to consider ordering processes in the subsystems of DMA groups. The interaction between groups in their various orientational states can then be examined using the dipole-dipole approximation. Next, the thermodynamic characteristics of the model can be calculated using the mean field approximation.

◎The conceptual model of groundwater flow for a particular site can be constructed. The conceptual model can then be transformed into a deterministic numerical model using MODFLOW. Additionally, a stochastic numerical model can be developed by linking the Kalman filter to the system and observation equations of the deterministic numerical model.

◎Geo-hydrological layers of existing and newly set wells can be analyzed and a numerical model can be developed as well Next, parameters can be estimated from the covariance of regional variables. Additionally, a 3D groundwater numerical simulation can verify the geo-hydrological analysis and confirm the aquifer's position. Moreover, with the geo-statistical method, the layered structure can be analyzed by the results of the groundwater numerical model. The statistical optimal monitoring design can be determined by variance reduction analysis.

◎Clement's method can be adopted to adjust the conventional indices. The bootstrap method can then be applied to reduce the estimation error. Next, computer simulations can evaluate different non-normal distribution processes to demonstrate the effectiveness of the proposed index. Additionally, a series of procedures can be developed for engineers without a statistical background.

◎A simulation model can be developed to distribute the releases from reservoirs based on the operating rules of multi-reservoirs. These releases can then be used to evaluate the agricultural and non-agricultural objectives. Finally, the non-inferior solution set, which reflects the impact on the relationship between agricultural and non-agricultural objectives, can be estimated by a multi-objective genetic algorithm.

◎Principal component analysis (PCA) can be performed to form new variables, which are the key components of original multiple characteristics in a manufacturing system. Their formation can decrease the number of control charts since PCA reduces the number of related features. A multivariate exponential weight moving average (MEWMA) control chart can then be used to verify whether the process is controlled. Additionally, for a situation in which MEWMA indicates that the process is out of control, three unique EWMA control charts of different sources of variation can be used to identify the source of inaccurate variation.

**3. Anticipated results 希望的結果.  What are the main results that you hope to obtain in your project? 你希望達成的結果?**

Consider the following examples.

◎As anticipated, the proposed model can adopt statistical methodologies to increase the availability of the IMD World Competitiveness Model by 20%.

◎As anticipated, the proposed response surface method can provide an optimization procedure with simplified calculations of the location and dispersion effects as well as to alleviate the problem of ordered categorical data quality.

◎As anticipated, the studied PCIs can be applied to investigate and evaluate the suppliers' DT and OCT, underlying the choice of suppliers.

◎For classroom use in distance learning courses, students can review the homework of their peers and receive comments. At the end of a semester, educators can easily access the student profiles via this system for further analysis.  Additionally, educators can identify not only the significant relationship between students' attitudes and their performance, but also the reliability and validity coefficients of networked peer assessment.

◎As anticipated, the proposed algorithm can reduce the computational time required to solve a nonlinear programming model by 50% of that required by piecewise linearization algorithms.

159

◎As anticipated, the novel architecture can automatically page PDA users through their handheld mobile devices when desired information becomes available. Moreover, the architecture allows PDA users to access information with their geographic position's functioning as a filter.

◎As anticipated, the novel web-based system in a PC-LAN environment can identify 85% of all network problems in 5 minutes by reducing the complexity of operating a network management system.

◎As anticipated, the proposed method can predict geometric parameters of a C4 type solder joint to within 5% of those obtained by a specific method found in the literature.

◎As anticipated, the proposed procedure can be used to estimate the marginal dimension of an order parameter, corresponding to experimental data found in pertinent literature. Several theoretical renormalization group methods can be adopted to estimate the marginal dimension, in addition to those available in the literature.

◎As anticipated, low-dose radiation processes can be clarified and the parameters of the II-VI barrier structures can be obtained .

◎As anticipated, the proposed stochastic model can provide further insight into the uncertainty of the estimation error of this model, by adopting different groundwater monitoring network alternatives.

◎As anticipated, the proposal model can accurately estimate the new sites of wells. These estimates can be used not only to design efficiently a ground water network, but also to enable decision makers to identify optimal positions.

◎As anticipated, the proposed index can effectively evaluate non-normal distribution processes.

◎As anticipated, the proposed model can determine when to transfer irrigation water from the irrigation association to non-agricultural areas, based on the non-inferior solutions, to offset water shortages during the dry season.

◎As anticipated, the control process can detect small shifts in a manufacturing system. Moreover, the control process can identify which sources of variation or characteristics are out of control.

**4. Contribution to field 領域的貢獻.  What is the main contribution of your project to the field or industrial sector to which your company or organization belongs? 你的提案對相關工作領域的貢獻?**

Consider the following examples.

◎The proposed approach can provide Taiwan with an effective strategy for enhancing its IMD world competitiveness ranking.

◎The reliability and validity of networked peer assessment can be demonstrated as a viable assessment strategy for distance learning.

◎The proposed editor, using a Java applet, can provide a graphical user interface and cross platform tools, allowing flexible operation and easy editing of mathematical symbols.

◎In addition to obtaining the global optimum in general nonlinear programming models to within a tolerable error, the proposed algorithm can significantly increase computational efficiency by decreasing the use of 0-1 variables.

◎The proposed model can minimize the errors associated with constructing a digital face, enhancing multimedia or animation applications by reducing formulation costs and creating more realistic digital objects.

◎The proposed method can reveal how thermodynamic functions depend on microscopic parameters near the binary symmetrical mixture critical points, clarifying how microscopic parameters influence macroscopic critical behavior.

◎Results of this study can provide further insight into the peculiarities of rebuilding radiation defects in cesium halides doped with OH-impurities.

◎The proposed method can be used to design geometric parameters of a C4 type solder joint, enhancing the reliability of a flip chip package and reducing its stress concentration.

◎The ability to determine how gamma radiation affects the formation of defects in CsJ crystals, doped with different impurities can hopefully lead to industrial applications.

◎The EEE mechanism for CsBr can be clarified, shedding further light on the formation of exoelectrons.

◎The proposed model can also accurately describe systems with more than two equilibrium positions on a site, extending conventional order-disorder models to a wider class of materials.

◎The proposed model can minimize the costs of constructing a monitoring network by assessing monitoring network alternatives.

◎In addition to minimizing the errors, the proposed model can reduce the number of monitoring wells.

◎Engineers can easily adopt the proposed index and related procedures when comparing processes or selecting an alternative supplier.

◎The proposed optimization model can provide a valuable reference for governmental authorities when drawing up water resource-related management strategies.

◎Results in this study can provide valuable manufacturing information to engineers who can quickly assess the conditions of a manufacturing system. By effectively responding to this information, engineers can promptly adjust the manufacturing system to enhance wafer quality.

**F** In the space provided, write a hypothesis statement for your work proposal.

Work objective 工作目標:

_____

_____

_____

Methodology to achieve objective 達成目標的方法:

_____

_____

_____

Anticipated results  希望的結果:

_____

_____

_____

Contribution to field  領域的貢獻:

_____

_____

_____

G Consider the following examples of how to write a hypothesis statement for a work proposal.

*Civil Engineering* 土木相關分類

**Work objective** 工作目標 A deterministic and stochastic model for simulating groundwater flow can be developed to assess monitoring network alternatives. **Methodology to achieve objective** 達成目標的方法 To do so, the conceptual model of groundwater flow for a particular site can be constructed.  The conceptual model can then be transformed into a deterministic numerical model using MODFLOW. Additionally, a stochastic numerical model can be developed by linking the Kalman filter to the system and observation equations of the deterministic numerical model. **Anticipated results** 希望的結果 As anticipated, the proposed stochastic model can provide further insight into the uncertainty of the estimation error of this model by adopting different groundwater monitoring network alternatives. **Contribution to field** 領域的貢獻 The proposed model can minimize the costs of constructing a monitoring network by assessing monitoring network alternatives.

*Civil Engineering* 土木相關分類

**Work objective** 工作目標 A groundwater numerical model can be developed to analyze the position and the quantities of optimal monitor wells in Choshuichi River Fan. **Methodology to achieve objective** 達成目標的方法 To do so, hydrological layers of existing and newly set wells can be analyzed and a numerical model can be developed. Next, parameters can be estimated from the covariance of regional variables. Additionally, a 3D groundwater numerical simulation can verify the geo-hydrological analysis and confirm the aquifer's position. Moreover, with the geo-statistical method, the layered structure can be analyzed by the results of the groundwater numerical model. The statistical optimal monitoring design can be determined by variance reduction analysis. **Anticipated results** 希望的結果 As anticipated, the proposal model can accurately estimate the new sites of wells. These estimates can be used not only to design a ground water network efficiently, but also to enable decision makers to identify optimal positions. **Contribution to field** 領域的貢獻 In addition to minimizing the tolerable errors, the proposed model can reduce the numbers of monitoring wells.

*Statistics* 統計相關分類

**Work objective** 工作目標 An appropriate process capability index can be developed to evaluate non-normal distribution processes. **Methodology to achieve objective** 達成目標的方法 To do so, Clement's method can be adopted to adjust the conventional indices. The Bootstrap method can then be applied to reduce the estimation error. Next, computer simulations can evaluate different non-normal distribution processes to demonstrate the effectiveness of the proposed index. Additionally, a series can be developed for engineers without a statistical background. **Anticipated results** 希望的 結果 As anticipated, the proposed index can effectively evaluate non-normal distribution processes. **Contribution to field** 領域的貢獻 Moreover, the proposed index and procedures can be easily adopted by engineers when comparing processes or selecting an alternative supplier.

*Civil Engineering* 土木相關分類

**Work objective** 工作目標 An effective water resource distributed optimization model can be developed not only to reflect accurately how multi-objectives compete with each other, but also to estimate the available releases of multi-reservoirs. **Methodology to achieve objective** 達成目標 的方法 To do so, a simulation model can be developed to distribute the releases among reservoirs based on the operating rules of multi-reservoirs. These releases can then be used to evaluate the agricultural and non-agricultural objectives. Finally, the non-inferior solution set which reflects the impact on the relationship between agricultural and non-agricultural objectives can be estimated by a multi-objective genetic algorithm. **Anticipated results** 希望的結果 As anticipated, the proposed model can determine when to transfer irrigation water from the irrigation association to non-agricultural areas, based on the non-inferior solutions, to offset water shortages during the dry season. **Contribution to field** 領域的貢獻 The proposed optimization model provides a valuable reference for governmental authorities when drawing up water resource-related management strategies.

*Industrial Engineering* 工業工程相關分類

**Work objective** 工作目標 A competent on-line control process can be developed for use in wafer fabs, capable of detecting assignable causes concealed behind both multiple characteristics and multiple readings in a manufacturing system with several sources of variation. **Methodology to achieve objective** 達成目標的方法 To do so, principal component analysis (PCA) can be performed to form new variables, which are the key components of original multiple characteristics in a manufacturing system. Their formation can decrease the number of control charts since PCA reduces the number of related features. A multivariate exponential weight moving average (MEWMA) control chart can then be used to verify whether the process is controlled. Additionally, for a situation in which MEWMA indicates that the process is out of control, three unique EWMA control charts of different sources of variation can be used to identify the inaccurate source of variation. **Anticipated results** 希望的結果As anticipated, the control process can detect small shifts in a manufacturing system. Moreover, the control process can identify which sources of variation or characteristics are out of control. **Contribution to field** 領域的貢獻 Results in this study can provide valuable manufacturing information to engineers  to quickly assess the conditions of a manufacturing system. By effectively responding to this information, engineers can promptly adjust the manufacturing system to enhance wafer quality.

*Statistics* 統計相關分類

**Work objective** 工作目標 An appropriate evaluation model capable of selecting natural gas bus brands can be developed. **Methodology to achieve objective** 達成目標的方法 To do so, natural gas bus brands can be selected via the proposed model by performing cost effectiveness analysis. Each criterion can also be evaluated under cost and effectiveness categories.  Two methodologies of multiple attribute decision making

(MADM), including the technique for order preference by similarity to ideal solution (TOPSIS) and the analytic hierarchy process (AHP), can then be used to rank all viable alternatives to bus systems from a complete perspective. **Anticipated results** 希望的結果 As anticipated, the ranking methodology can provide a more objective outcome with weights of related decision groups, than can other methodologies. The ranking methodology can also provide a more flexible procedure with respect to the outcome's complexity. Moreover, the ranking outcome can allow decision makers to identify the order preference for alternatives. **Contribution to field** 領域的 貢獻 The proposed model can also evaluate the relationship between the cost and effectiveness of all viable alternatives to bus systems. Furthermore, the proposed model can provide a valuable reference for government when selecting brands of bus systems.

*Statistics* 統計相關分類

**Work objective** 工作目標 A novel response surface methodology can be developed to optimize efficiently ordered categorical data process parameters since setting the process parameters leads to optimization of the location and dispersion effects. **Methodology to achieve objective** 達成目 標的方法 To do so, this novel methodology with simplified calculations of the mean and standard deviation of ordered categorical responses can be used to estimate the location and dispersion effects. Additionally, regression models can be used to relate the location and dispersion effects to the controlled factor levels. Finally, an optimal combination of process parameters can be obtained using the dual response surface method. **Anticipated results** 希望的結果 As anticipated, the proposed response surface method can provide an optimization procedure with simplified calculations of the location and dispersion effects and alleviate the problems related to ordered categorical data quality.

*Industrial Engineering* 工業工程相關分類

**Work objective** 工作目標 An efficient hypothesis testing procedure for PCIs can be developed, capable of assessing the operational cycle time (OCT) and delivery time (DT) for VLSI. **Methodology to achieve objective** 達成目標的方法 To do so, the PCIs of OCT and DT can be defined, and the UMVU (uniformly minimum variance unbiased) estimators of the studied PCIs can be derived under the assumption of a normal distribution. The above estimators can then be used to construct a one-to-one relationship between the PCIs and the conforming rate of DT (or OCT). Finally, a hypothesis testing procedure for PCIs can be developed. **Anticipated results** 希望的結果 As anticipated, the studied PCIs can be applied to investigate and evaluate the suppliers' DT and OCT, underlying the choice of suppliers. **Contribution to field** 領域的貢獻 Moreover, PCIs can be used in uniform standards to assess DT and OCT for VLSI.

*Distance Learning* 遠距教學相關分類

**Work objective** 工作目標 An Internet-based constructive learning environment can be developed so that participants can interactively link their concept maps for accumulative learning. **Methodology to achieve objective** 達成目標的方法 To do so, three groups (each comprised of 10 to 11 students) can be formed, and a chapter assigned to each group for reading and concept mapping. Instructed to construct individual concept maps in a knowledge construction environment, each student can only view his or her own concept map to prevent students from copying their peers. Next, after reading concept maps constructed by the other groups, students can select the best concept map. Students can be instructed to link their concept maps to the best one. The concept maps of each student can then be evaluated by voting. **Anticipated results** 希望的結果 As anticipated, using this approach, the individual construction stage and the interlinking stage can together form the cooperative-competitive model. The Internet-based learning environment can also prevent unequal participation in group activities. **Contribution to field** 領域的貢獻 Moreover, this environment can provide a flexible cooperative-competitive model for educators to use in classroom activities.

*Distance Learning* 遠距教學相關分類

**Work objective** 工作目標 A networked peer assessment system capable of supporting instruction and learning can be developed to analyze students' learning outcomes in higher education and collect valuable student profiles for analysis by educators. **Anticipated results** 希望的結果 For classroom use in distance learning courses, students can review the homework of their peers and receive comments. At the end of a semester, educators can easily access the student profiles via this system for further analysis. Additionally, educators can identify not only the significant relationship between students' attitudes and their performance, but also the reliability and validity coefficients of networked peer assessment. **Contribution to field** 領域的貢獻 Moreover, the reliability and validity of networked peer assessment can be demonstrated as a viable assessment strategy for distance learning.

*Computer Science* 資訊科學相關分類

**Work objective** 工作目標 A user-friendly editor of mathematical symbols with a graphic user interface can be developed using a Java applet. **Anticipated results** 希望的結果 As anticipated, the proposed editor can run on all Java-enabled web browsers, regardless of the platform. **Contribution to field** 領域的貢獻 The proposed editor, using a Java applet, can provide a graphical user interface and cross platform tools, allowing flexible operation and easy editing of mathematical symbols.

*Information Management* 資訊管理相關分類

**Work objective** 工作目標 Combining the flexibility of a network with the storage capacity of a computer, a networked portfolio system using peer assessment can be developed to assess the higher-order thinking of students. **Methodology to achieve objective** 達成目標的方法Based on portfolio and peer assessment, the networked portfolio system can provide an environment that combines the flexibility of a network with the storage capacity of a computer. This system can also allow students to collect their learning recode (including their homework), interact with peers, and critically reflect. **Anticipated results** 希望的結果 As anticipated, in addition to helping teachers and researchers assess the higher-order thinking of students, the proposed system can also help students to hone their critical thinking and analytical skills. **Contribution to field** 領域的貢獻 The networked electronic portfolio system and peer assessment can be used at all educational levels.

*Computer Science* 資訊科學相關分類

**Work objective** 工作目標 An enhanced piecewise linearization algorithm, capable of obtaining the global optimum of a nonlinear model, can be developed for use in a web based optimization system. **Methodology to achieve objective** 達成目標的方法 To do so, a web-based optimization system can be implemented based on the enhanced algorithm and using a dynamic linking library procedure. The system can then be linked to many other mathematical methods, for example, LINGO, to solve a nonlinear problem by integrating concurrent methods. Next, user specified problems can be stored in a database storage system. Additionally, the solution can be derived to guarantee the global optimum with an acceptable error rate. **Anticipated results** 希望的結果 As anticipated, the proposed algorithm can reduce the computational time required to solve a nonlinear programming model to 50% of that required by piecewise linearization algorithms. **Contribution to field** 領域的貢獻 Moreover, in addition to obtaining the global optimum in general nonlinear programming models to within a tolerable error, the proposed algorithm can significantly increase computational efficiency by decreasing the use of 0-1 variables.

*Information Management* 資訊管理相關分類

**Work objective** 工作目標 An efficient face model can be developed, capable of formulating the 3D image of an individual's face from three 2D images. **Methodology to achieve objective** 達成目標的方法 To do so, three 2D images can be obtained simultaneously with three general cameras. The images can then be transfered to a personal computer. Next, the proposed model can be used to formulate digitally the 3D face. **Anticipated results** 希望的結果 As anticipated, the proposed model can reduce the time to construct a 3D face by 10%, by optimizing the 3D model rather than manually retrieving 2D images. **Contribution to field** 領域的

貢獻 Moreover, the proposed model can minimize the tolerable errors associated with constructing a digital face, enhancing multimedia or animation applications by reducing formulation costs and creating more realistic digital objects.

*Information Management* 資訊管理相關分類

**Work objective**工作目標 A GIS-based architecture that supports an automatic reporting service through handheld mobile devices can be designed. **Methodology to achieve objective** 達成目標的方法 To do so, a GPS module can be used to obtain the coordinates of the PDA user. The coordinates can then be transmitted to the back-end server through a wireless network and used as the filter to query the database. Query results can then be sent back to the PDA, triggering an event to inform the PDA user. **Anticipated results** 希望的結果 As anticipated, the novel architecture can automatically page PDA users through their handheld mobile devices when desired information becomes available. Moreover, the architecture allows PDA users to access information with their geographic position's functioning as a filter. **Contribution to field** 領域的貢獻 The architecture proposed herein can hopefully be commercialized in PDA software.

*Information Management* 資訊管理相關分類

**Work objective** 工作目標 A novel web-based system in a PC-LAN environment capable of detecting network problems can be developed. **Methodology to achieve objective** 達成目標的方法 To do so, a graphic interface can be designed to illustrate the network structure of an enterprise. A free software program known as Multi Router Traffic Grapher (MRTG) can then be applied to communicate with network devices. Return values from network devices can then be filtered to detect potential network problems. Additionally, the visualized information can be updated on the user interface of a web page. **Anticipated results** 希望的結果 As anticipated, the novel web-based system in a PC-LAN environment can identify 85% of all network problems in 5 minutes by reducing the complexity of operating a network management system. **Contribution to field** 領域的貢獻 Moreover, the proposed system can be used to ensure flexible and inexpensive network management for enterprises, by combining free software from the Internet.

*Computer Science* 資訊科學相關分類

**Work objective** 工作目標 A novel learning environment can be developed, capable of assisting students in flexibly learning genetic algorithms based on computer-assisted instruction. **Methodology to achieve objective** 達成目標的方法 To do so, several benchmark problems can be integrated in this environment. A mathematical expression can then be developed to provide users with fitness functions of GAs. Next, a case study involving a GAs course can be presented to demonstrate the effectiveness of the proposed environment. **Anticipated results** 希望的結果 As anticipated, the proposed environment can reduce the time required for students to complete a GA assignment to one week, increasing the number of practice exercises that can be implemented and allowing them to better learn Gas. **Contribution to field** 領域的貢獻 Moreover, the learning environment can eliminate the need for hand coding GA programs, simplifying the process of learning genetic algorithms.

*Computer Science* 資訊科學相關分類

**Work objective** 工作目標 An automatic knowledge-base construction methodology, in which the knowledge bases are represented by fuzzy concept networks, can be developed in three stages. **Anticipated results** 希望的結果 As anticipated, the proposed methodology can reduce the time for constructing the knowledge base by 80%. **Contribution to field** 領域的貢獻 Therefore, the proposed methodology can be adopted to construct a knowledge base in an intelligent information retrieval system more efficiently.

*Mechanical Engineering* 機械相關分類

**Work objective** 工作目標 An analytical geometry method can be developed, capable of accurately predicting the geometric parameters of practical C4 type solder joint in flip chip technology after a reflow process. **Methodology to achieve objective** 達成目標的方法 To do so, the differential equation of curvature can be derived from the well-known Laplace-Young equation. Additionally, the contact angle between the solder and substrate and the surface tension of the solder can be measured. The free surface energy of the substrate and the interfacial tension between the solder and the substrate can then be obtained from Young's equation. Next, the differential equation of curvature can be numerically solved to obtain the geometric parameters. The geometric data of the solder joint can be further used as an input in the finite element model to analyze the stress/strain distribution, thermal fatigue life and reliability of the electronic packages. **Anticipated results** 希望的結果 As anticipated, the proposed method can predict geometric parameters of a C4 type solder joint to within an accuracy 5% of those obtained by a specific method found in the literature. **Contribution to field** 領域的貢獻 Moreover, the proposed method can be used to design geometric parameters of a C4 type solder joint, capable of enhancing the reliability of a flip chip package and reducing its stress concentration.

*Computer Science* 資訊科學相關分類

**Work objective** 工作目標 An effective GA model can be developed, capable of deriving an optimal solution for each state in a multi-state problem. **Methodology to achieve objective** 達成目標的方法 To do so, the proposed fuzzy polyploidy, a multi-state chromosome coding scheme, can be used to describe the solution of a multi-state problem. Moreover, an adaptive genetic structure model can be adopted to derive an appropriate polyploidy structure for practical applications. The proposed model consists of three structural level operations, including structural expansion, structural deletion, and structural coercion, to simulate the natural random variation. **Anticipated results** 希望的結果 As anticipated, the proposed GA model can increase the accuracy of the optimum solution derived for a particular multi-state problem. **Contribution to field** 領域的貢獻 Moreover, the proposed model can enhance conventional genetic algorithms by systematically solving multi-state problems through the use of the polyploidy concept.

*Physics* 物理相關分類

**Work objective** 工作目標 An estimation method capable of obtaining the marginal number of order parameter dimensions can be developed. **Methodology to achieve objective** 達成目標的方法 To do so, perturbation theory can be expanded for mc. The resummation procedure can then be applied to those expansions. Next, resummation results can be analyzed to estimate the marginal dimension of a weakly diluted O(m) model. **Anticipated results** 希望的結果 As anticipated, the proposed procedure can be used to estimate the marginal dimension of an order parameter, corresponding to experimental data found in pertinent literature. Several theoretical renormalization group methods can be adopted to estimate the marginal dimension, in addition to those available in the literature, **Contribution to field** 領域的貢獻 yielding accurate results for a weakly diluted O (m) model that can thoroughly elucidate its critical behavior.

*Physics* 物理相關分類

**Work objective** 工作目標 The critical behavior of a binary symmetrical mixture can be elucidated using the collective variables method. **Methodology to achieve objective** 達成目標的方法 To do so, the grand partition function of a binary symmetrical mixture can be constructed near the vapour-liquid and mixing-demixing critical point. The grand partition function can then be reduced to the partition function of a three-dimensional Ising model in an external field. Next, for a hard-sphere square-well symmetrical binary mixture, the thermodynamic characteristics can be calculated as a function of microscopic parameters. **Anticipated results** 希望的結果 As anticipated, adopting the collective variables method allows the non-universal critical characteristics of a hard-sphere square-well binary symmetrical mixture to be quantified. **Contribution to field** 領域的貢獻 Moreover, the proposed method can reveal how thermodynamic functions depend on microscopic parameters in near the binary symmetrical mixture critical points, clarifying how microscopic parameters influence macroscopic critical behavior.

*Physics* 物理相關分類

**Work objective** 工作目標 Cesium iodine and cesium bromide crystals doped with OH impurities can be grown by different methods, including Stockbarger in vacuum and Kyropolous in air. **Methodology to achieve objective** 達成目標的方法 To do so, gamma-irradiation can be performed at room temperature using a Co60 gamma-source at a dose rate of 3800 rad.s-1. Next, conductivity versus temperature can be measured for irradiated and non-irradiated crystals. **Anticipated results** 希望的結果 As anticipated, the conductivity can be explained by the different thermal stabilities of simple and complicated radiation defects and by accompanying recombination processes. **Contribution to field** 領域的貢獻 Results in this study can provide further insight into the peculiarities of rebuilding radiation defects in cesium halides doped with OH-impurities.

*Physics* 物理相關分類

**Work objective** 工作目標 Dependence of efficiency of formation efficiency of defects in CsJ crystals on both types of impurity and on different values of absorbed radiation dose can be investigated. **Methodology to achieve objective** 達成目標的方法 To do so, the dipole relaxation processes in doped CsJ crystals can be elucidated. **Anticipated results** 希望的結果 As anticipated, the mechanism of defect formation and decay can be ascertained. **Contribution to field** 領域的貢獻 The ability to determine how gamma irradiation affects the formation of defects in CsJ crystals doped with different impurities can hopefully lead to industrial applications.

*Physics* 物理相關分類

**Work objective** 工作目標 A novel order-disorder four-state model can be developed, capable of theoretically describing the dielectric properties of the DMAGaS-DMAAlS ferroelectric crystals. **Methodology to achieve objective** 達成目標的方法 To do so, the model Hamiltonian can be constructed to consider ordering processes in the subsystems of DMA groups. The interaction between groups in their various orientational states can then be examined using the dipole-dipole approximation. Next, the thermodynamic characteristics of the model can be calculated using the mean field approximation. **Anticipated results** 希望的結果 As anticipated, the proposed model can clarify the dielectric phenomena during the transition and disappearance of the ferroelectric phase under hydrostatic pressure. **Contribution to field** 領域的貢獻 The proposed model can also accurately describe systems with more than two equilibrium positions on a site, extending conventional order-disorder models to a wider class of materials.

*Physics* 物理相關分類

**Work objective** 工作目標 The RSOE mechanism can be investigated and, based on that data, a theoretical model can be constructed. **Methodology to achieve objective** 達成目標的方法 To do so, the preliminary experimental data can be obtained by measuring voltage-current characteristics (VIC), voltage-capacity characteristics (VCC) and capacity-modulated spectra for barrier structures at different irradiation doses. The required parameters can then be derived from the obtained data. Additionally, Hall investigations can be undertaken to consider bulk effects. The changes of parameters can be analyzed and synthesized. **Anticipated results** 希望的結果 As anticipated, low-dose radiation processes can be clarified and the parameters of the II-VI barrier structures can be obtained. **Contribution to field** 領域的貢獻 Thus, the range/number of the RSOE objects can be increased and the RSOE mechanisms can be distinguished.

*Physics* 物理相關分類

**Work objective** 工作目標 Whether the defect-recombination mechanism is responsible for CsBr EEE can be investigated. **Methodology to achieve objective** 達成目標的方法 To do so, a wave function can be chosen for the initial and final state of the exoelectron. The probability of transition of the electron from the F- center state to the free state can then be theoretically calculated. Next, the energy distribution parameters can be obtained. Moreover, the theoretical results can be compared to the experimental results. **Anticipated results** 希望的結果 As anticipated, the energy spectra of the exoelectrons can be obtained from CsBr. **Contribution to field** 領域

的貢獻 Moreover, the EEE mechanism for CsBr can be clarified, shedding further light on the formation of exoelectrons .

*Mechanical Engineering* 機械相關分類

**Work objective** 工作目標 An SAR-controlled DLL deskew circuit capable of reducing the system clock skew problem can be developed. **Methodology to achieve objective** 達成目標的方法 To do so, SAR can be used to control the DLL so that the deskew circuit can be automatically optimized for clock synchronization. The SAR binary search method can then be employed to reduce the lock time and maintain tight synchronization. Moreover, with the clock-deskew buffers' using inverter chains, the deskew circuit can reduce the system clock skew and obtain a perfect output clock duty cycle. Furthermore, the delay buffer chain can be adjusted so that the deskew circuit can fit in different operating environments. Finally, the architecture can be implemented using Synopsys and Cadence tools. **Anticipated results** 希望的結果 As anticipated, the proposed circuit can synchronize the 100~133 MHz system clock within 21 clock cycles. **Contribution to field** 領域的貢獻 Moreover, the proposed circuit can eliminate the system malfunction caused by the clock skew problem.

*Information Management* 資訊管理相關分類

**Work objective** 工作目標 A novel competitiveness model capable of integrating a global optimization algorithm into the IMD World Ranking can be developed. **Methodology to achieve objective** 達成目標的方法 To do so, populations can be collected from data of the IMD World Competitiveness Yearbook. Variables can then be analyzed. Next, the populations can be grouped. **Anticipated results**希望的結果 As anticipated, the proposed model can adopt statistical methodologies to increase the availability of the IMD World Competitiveness Model by 20%. **Contribution to field** 領域的貢獻 Also, the proposed approach can provide Taiwan with an effective strategy for enhancing its IMD world competitiveness ranking.

*Electrical Engineering* 電子相關分類

**Work objective** 工作目標 a program can be developed for different blocks in the chipset to verify whether a test pattern is correct. **Methodology to achieve objective** 達成目標的方法 To do so, limitations and properties of the tester can be reviewed. The proposed program can then be used to verify the testing patterns. If errors in those patterns are detected, corrections can be made without using the tester to verify, saving considerable time. **Anticipated results** 希望的結果 As anticipated, the proposed strategy allows testers to examine the test pattern more precisely than do conventional approaches. **Contribution to field** 領域的貢獻 The proposed strategy can reduce the inspection time by writing a rule-check program and familiarizing the user with related software.

*Electrical Engineering* 電子相關分類

**Work objective** 工作目標 A numerical method can be developed to choose efficiently the bandwidth of the loop filter in the PLL circuit. An additional apparatus can also be developed in the carrier recovery circuit to estimate offsets precisely. **Methodology to achieve objective** 達成目標的 方法 To do so, a frequency detection apparatus can be used in the carrier recovery circuit to lock the large frequency offset. The PLL circuit can then automatically switch the coefficients of the loop filter into distinct bandwidths to reduce vibration and to converge faster than conventional circuits. **Anticipated results** 希望的結果 As anticipated, the novel design can lock a wide range of offsets of more than 100 KHz in a short acquisition time with a symbol error rate of less than 0.01. **Contribution to field** 領域的貢獻 The improved carrier recovery on a digital receiver can track better than conventional models, making telecommunication products more competitive.

*Electrical Engineering* 電子相關分類

**Work objective** 工作目標 A low re-deposition process in HDP CVD can be developed to increase the gap-filling capability of the next IC manufacturing generation. **Methodology to achieve objective** 達成目標的 方法 To do so, the process pressure can be decreased by reducing the deposited gas flow. After gas sputtering is decreased, the RF power and gas ratio can be adjusted. **Anticipated results** 希望的結果 As anticipated, the proposed process can achieve an average aspect ratio of 2.7 for 0.2 um metal spacing. **Contribution to field** 領域的貢獻 By significantly enhancing the gap filling capacity to satisfy 0.15 um process requirements, the HDP-CVD process proposed herein can provide a good solution for metal interconnection process when shrinking the dice size in next generation.

*Electrical Engineering* 電子相關分類

**Work objective** 工作目標 A novel connecting interface for hard discs can be developed for supporting up to 150 MHz bandwidths and minimizing the number of pins to four. **Methodology to achieve objective** 達成目標的方法 To do so, this novel connecting interface with serial transmitting and receiving lines can be used to facilitate communication between system and hard disc. The serial interfaces can then be used to eliminate the cross talk between transmitted and received data, discovered in conventional interfaces, and can easily support bandwidths of up to 150 MHz. Next, the minimal pin requirement can be satisfied using the serial interfaces. **Anticipated results** 希望的結果 As anticipated, the novel serial interface can facilitate robust communication between system and hard disc and easily support up to 150 MHz bandwidths. Moreover, the proposed interface can allow chipset vendors to integrate more functions into one chip than they did before. **Contribution to field** 領域的貢獻 In addition to making a high performance hard disc commercially feasible, the proposed interface can be easily upgraded in the future.

*Finance* 財務相關分類

**Work objective** 工作目標 A precise dynamic capacity model can be developed, capable of accurately forecasting capacity. **Methodology to achieve objective** 達成目標的方法 To do so, parametric effectiveness analysis can be performed to forecast capacity accurately from the input data. Each criterion can also be evaluated under quantity and time phase categories, using historical data. The parameters can be measured as well. Additionally, the wafer start schedule and equipment status (which are the primary parameters) can then be regulated to rank all viable alternatives to forecast the capacity. **Anticipated results** 希望的結果 As anticipated, the ranked forecasting capacities can provide a more objective outcome, with the weights of the Marketing Department, than can conventional models. Moreover, the ranked forecasting capacities allow the Sales Department to identify the influence of customers' orders on alternatives. **Contribution to field** 領域的貢獻 The dynamic capacity model can also evaluate the cost and effectiveness of all viable alternatives to our fabrication, in which the quantity of work in process influences our inventory costs.

*Electrical Engineering* 電子相關分類

**Work objective** 工作目標 Effective test parameters can be developed to demonstrate their utility in wafer testing. **Methodology to achieve objective** 達成目標的方法 To do so, pertinent literature and relevant industrial specifications can be surveyed. Engineering experiments can then be performed to verify the hypothesis. A monitor pin can also be added to take accurate measurements. **Anticipated results** 希望的結果 As anticipated, the proposed test parameters can reveal that knowing more exactly the shortage of test items reduces the test time and increases product quality. **Contribution to field** 領域的貢獻 The proposed test parameters can be used to establish a productive qualification flow, increasing capacity and reliability.

*Electrical Engineering* 電子相關分類

**Work objective** 工作目標 An emulation method, costing less and performing more efficiently than conventional methods, can be developed. **Methodology to achieve objective** 達成目標的方法 To do so, appropriate types of FPGA, for example, Virtex E of Xilinx or APEX of Altera, with a retail value of around US$10,000, can be selected according to size and speed requirements, to emulate an actual chip. An FPGA synthesis tool, for example, FPGA compiler or Synplify PRO, can then be adopted to map the design net-list into a edif file for FPGA. Next, software for placing and routing FPGA can be employed to construct an emulation database in FPGA. Additionally, timing constraints can be established to enhance the emulation. Moreover, the database can be downloaded into FPGA to initiate emulation and debug new designs **Anticipated results** 希望的結果 As anticipated, the proposed method can reduce the costs and time of developing a 10MHz frequency by 95%, whereas the conventional method can only be used at a frequency of 1MHz. **Contribution to field** 領域的貢獻 The proposed method can reduce testing time, include new devices that cannot function at 1MHz, and allow our company to verify the design with numerous platforms.

*Finance* 財務相關分類

**Work objective** 工作目標 A novel numerical model can be developed, capable of directly deriving a reasonable forward rate curve in one explicit function to correlate well with market data. **Methodology to achieve objective** 達成目標的方法 To do so, when including observable points, the forward rate curve of the proposed model can be expressed as a function of a specified form with a finite number of parameters. The maximum smoothness term can then be found within this parametric family to derive the model. **Anticipated results** 希望的結果 As anticipated, the proposed methods can provide the smoothest possible forward rate curve which is consistent with the chosen functional form and correlates with all observed market data on the yield curve. **Contribution to field** 領域的貢獻 Moreover, the proposed model can also provide an innovative means of directly fitting the forward rate with maximum smoothness and precision, providing a valuable solution for pricing interest rate derivatives.

*Electrical Engineering* 電子相關分類

**Work objective** 工作目標A robust and concise architecture can be developed to satisfy the quality and performance requirements of a 3D driver. **Methodology to achieve objective** 達成目標的方法 To do so, the hardware acceleration capabilities of the ICD architecture developed by SGI and Microsoft can be adopted to meet industrial standards. The driver can then be developed by incorporating user friendly properties. **Anticipated results** 希望的結果 As anticipated, our product can compete with other commercially available products and can generate large revenues. **Contribution to field** 領域的貢獻 The proposed driver can also ensure flexible use by the end user.

*Chemistry* 化學相關分類

**Work objective** 工作目標 A low impurity ( 3N ) NF3 clean gas can be developed, capable of reducing the costs of the DCVD clean process. **Methodology to achieve objective** 達成目標的方法 To do so, the clean test rate can be used to evaluate the 3N NF3 gas cleaning efficiency. The dielectric film's contamination test can then be performed by ICP-MS and VPD-TXRF. **Anticipated results** 希望的結果 As anticipated, according to the clean test rate, the clean efficiency of the 3N NF3 gas can be comparable to that of the high impurity gas. According to the dielectric film's contamination test, the cleaning efficiency of the 3N NF3 gas can be comparable to that of both clean gases. However, the end-point time of the cleaning process for both clean gases is the same. **Contribution to field** 領域的貢獻 The low impurity ( 3N ) NF3 clean gas can reduce the costs of the DCVD cleaning process by more than 50%.

# *Unit Five*

 Richard　Teacher Education(師資教育)

 Ann　Computer Science(資訊科學)

Novel learning environment

## Writing the abstract (part one): briefly introducing the background, objective and methodology

### 摘要撰寫（第一部分）：簡介背景、目標及方法

**Vocabulary and related expressions　相關字詞**

Taiwan's Ministry of Education 台灣的教育部長

multi-assessment 多向評估

assess 評估

higher-order thinking 高階思考

motivate students properly 正確地激發學生

flexibility 彈性

storage capacity 貯藏量

novel 新的

networked portfolio system 網路組合系統

peer assessment 同儕互評

Genetic Algorithm (GA) courses 基因演算法課程

optimization problems 問題的最佳化

coding programs 撰寫程式

implemented 工具

a novel learning environment 新的學習環境

computer-assisted instruction 電腦輔助教學

benchmark problems 可作為比較標準的題目

mathematical expression 數學上的表達

fitness functions 效益函數

case study 範例研究

Taiwan's global competitiveness ranking 台灣的全球競爭力等級

IMD World Competitiveness Scoreboard 洛桑管理學院世界競爭力排行榜

simulations 模擬

neglect 忽視

policy makers 決策者

competitiveness model 競爭力模式

a global optimization algorithm 全球最佳化演算法

population data 人口資料

variables 變量

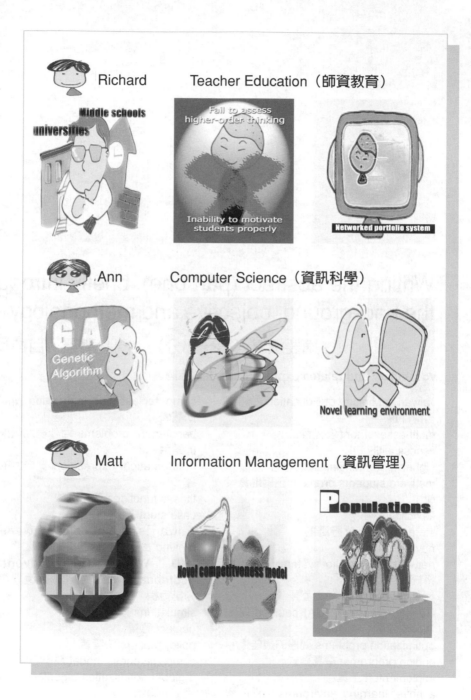

**A** Write down the key points of the situations on the preceding page while the instructor reads aloud the script on page 336.

**Situation 1**

_____

_____

_____

**Situation 2**

_____

_____

_____

**Situation 3**

_____

_____

_____

*Unit*
*Five*
Writing the abstract (part one): briefly introducing the
background, objective and methodology
摘要撰寫（第一部分）：簡介背景、目標及方法

**B** Based on the three situations in this unit, write three questions with replies of either *Yes* or *No*.

**Examples**

*Is Taiwan's global competitiveness ranking in the IMD World Competitiveness Scoreboard rising?*

*No, it is falling.*

*Can Matt's competitiveness model integrate a global optimization algorithm into the IMD World Competitiveness Model?*

*Yes*

1. _____

   _____

2. _____

   _____

3. _____

   _____

**C** Based on the three situations in this unit, write three questions beginning with *What*, and answer them.

**Examples**

*What is Taiwan's Ministry of Education increasingly emphasizing?*

*The use of multi-assessment in middle schools and universities*

*What is the main feature of Matt's competitiveness model?*

*It can integrate a global optimization algorithm into the IMD World Competitiveness Model.*

1. _____

   _____

2. _____

   _____

3. _____

*Unit*
*Five*

Writing the abstract (part one): briefly introducing the
background, objective and methodology
摘要撰寫（第一部分）：簡介背景、目標及方法

**D** Based on the three situations in this unit, write three questions beginning with *Why*, and answer them.

**Examples**

*Why do conventional methods of assessing students' abilities fail to assess higher-order thinking?*

*Because they can not motivate students properly*

*Why does Ann want to develop a mathematical expression?*

*To provide fitness functions of GAs*

1. _____

_____

2. _____

_____

3. _____

_____

## E Write questions that match the answers provided.

**Examples**

*What do the IMD Competitiveness Simulations neglect?*

*Interactions among factors*

*Why does Ann present a case study of a GA course?*

*To demonstrate the effectiveness of the proposed environment*

1. _____

_____

Assisting students in flexibly learning genetic algorithms

*Unit*   Writing the abstract (part one): briefly introducing the
*Five*   background, objective and methodology
        摘要撰寫（第一部分）：簡介背景、目標及方法

2. _____

   _____

   To provide fitness functions of GAs

3. _____

   _____

   Coding programs as exercises

# Unit Five

Writing the abstract（part one）：
briefly introducing the background,
objective and methodology

## 摘要撰寫（第一部分）：簡介背景、目標及方法

1 Setting of work proposal 工作提案建構
2 Work problem 工作問題
3 Work objective 工作目標
4 Methodology to achieve objective 達成目標的方法

*Unit*
*Five*
Writing the abstract (part one): briefly introducing the
background, objective and methodology
摘要撰寫（第一部分）：簡介背景、目標及方法

Writing the first part of the abstract involves the following parts.

**1. Setting of work proposal 工作提案建構. What is the topic with which your work proposal is concerned? Can you point the reader of your work proposal to the general area with which you are concerned? 你工作提案的主題是什麼? 你的讀者可以明瞭工作提案的內容嗎?**

Consider the following examples.

◎An increasing number of Genetic Algorithm (GA) courses are offered to solve optimization problems.

◎Weak disorder leads to two critical behavior scenarios of the O(m) model.

◎The interaction between binary mixture components leads to diverse phase behavior with respect to relative molecular sizes and the strengths of their interactions.

◎As well known, radiation induces defect rebuilding.

◎Dielectric properties of DMAGaS-DMAAlS ferroelectrics produce variations temperature and pressure behavior.

◎The surface control method based on exoelectron emission (EEE), a precise and nondestructive relaxational method, can detect and predict the early stages of the destruction of materials. Additionally, the exoemission from alkali halide crystals has been extensively studied theoretically and experimentally.

◎Evaporation methods enhance the formation of scattering centers and the electrical conductivity of thin metal films.

◎Dipole relaxation arises in doped CsJ crystals.

◎Taiwan's global competitiveness ranking in the IMD World Competitiveness Scoreboard is falling.

**2. Work problem 工作問題. What is the problem that you are trying either to solve or more thoroughly to understand in the work proposal? 你的工作提案裡有你試著要解決或是想更進一步瞭解的問題嗎?**

Consider the following examples.

◎Activities of many distance learning environments over rely on in-group cooperation, erroneously implying that group members participate equally.

◎Conventional methods of assessing students' abilities fail to assess higher-order thinking owing to their inability to motivate students properly.

◎Recent studies have suggested that many web-learning activities merely support the memorization of facts, passive learning, or even disenchanted browsing.

◎Piecewise linearization algorithms require too much time to obtain an optimum solution.

◎Conventional 3D models are too time consuming and inaccurate when constructing digital objects since they manually retrieve 2D images.

◎The passive mode of accessing information fails to transmit effectively geographic-based information to PDA users.

◎Most knowledge bases are constructed by inquiring of domain experts to acquire knowledge, which requires significant time.

◎Geometric parameters of a C4 type solder joint can not be accurately predicted.

**3. Work objective 工作目標. What is the objective of your work proposal? 工作提案的目標？**

Consider the following examples.

◎This work presents a networked electronic portfolio system with peer assessment to provide a creative means of assessing the higher-order thinking of students.

◎This study analyzes the online discussions of collaborative teams to reveal whether such activity engages students in higher order thinking and how it takes place.

◎This work proposes an enhanced piecewise linearization algorithm, capable of obtaining the global optimum of a nonlinear model, for use in a web based optimization system.

◎This work describes an efficient face model capable of formulating the 3D image of an individual's face from three 2D images.

◎This work presents a novel web-based system in a PC-LAN environment capable of detecting network problems.

◎This study describes a novel learning environment capable of assisting students in flexibly learning genetic algorithms based on computer-assisted instruction.

◎This work develops an analytical geometry method capable of accurately predicting the geometric parameters of a C4 type solder joint in flip chip technology after a reflow process.

◎This work designs an effective GAs model capable of deriving an optimal solution for each state in a multi-state problem.

◎This study presents a novel estimation method capable of obtaining the marginal number of order parameter dimensions.

◎This investigation elucidates the critical behavior of a binary symmetrical mixture using the collective variables method.

◎This study develops a novel order-disorder four-state model capable of theoretically describing the dielectric properties of DMAGaS-DMAAIS ferroelectric crystals.

◎This study investigates the role of RSOE in II-VI semiconductors and constructs a related model.

**4. Methodology to achieve objective 達成目標的方法．What are the steps in your project to achieve the above objective? 你的計畫中達成目標的步驟?**

Consider the following examples.

◎The PCIs of OCT and DT are defined, and the UMVU (uniformly minimum variance) estimators of the studied PCIs are derived under the assumption of a normal distribution. These estimators are then used to construct the one-to-one relationship between the PCIs and the conforming rate of DT (or OCT). Finally, a hypothesis testing procedure for PCIs is developed.

◎A web-based optimization system is implemented based on the enhanced algorithm and using a dynamic linking library procedure. The system is then linked to many other mathematical methods, for example, LINGO, to solve a nonlinear problem by integrating concurrent methods. Next, user specified problems are stored in a database storage system. Additionally, the solution is derived to guarantee the global optimum with an acceptable error rate.

◎A GPS module is used to obtain the coordinates of the PDA user. The coordinates are then transmitted to the back-end server through a wireless network and used as the filter to query the database. Next, query results are sent back to the PDA, triggering an event to inform the PDA user.

*Unit*
*Five*
Writing the abstract (part one): briefly introducing the
background, objective and methodology
摘要撰寫（第一部分）：簡介背景、目標及方法

◎A graphic interface is designed to illustrate the network structure of an enterprise. A free software program known as Multi Router Traffic Grapher (MRTG) is then applied to communicate with network devices. Return values from network devices are then filtered to detect potential network problems. Addiitonally, the visualized information is updated on the user interface of a web page.

◎The differential equation of curvature is derived from the well-known Laplace-Young equation. Additionally, the contact angle between the solder and substrate and the surface tension of the solder can be measured. The free surface energy of the substrate and the interfacial tension between the solder and the substrate can then be obtained from Young's equation. Next, the differential equation of curvature is numerically solved to obtain the geometric parameters. Finally, the geometric data of the solder joint is used as an input in the finite element model to analyze the stress/strain distribution, thermal fatigue life and reliability of the electronic packages.

◎Voltage-current characteristics (VIC), voltage-capacity characteristic (VCC) and capacity-modulated spectra for barrier structures are measured at various irradiation doses, thus providing the preliminary experimental data. Based on that data, the control parameters are then derived. Following Hall experiments to consider bulk effects, the parametric changes are analyzed and synthesized.

◎SAR is used to control the DLL so that the deskew circuit can automatically optimize the delay time for clock synchronization. The SAR binary search method is then employed to reduce the lock time and maintain tight synchronization. Moreover, with the clock-deskew buffers' using inverter chains, the deskew circuit can reduce the system clock skew and obtain a perfect output clock duty cycle. Furthermore, the delay buffer chain is adjusted so that the deskew circuit can fit in different operating environments. Finally, the architecture is implemented using Synopsys and Cadence tools.

206

**F** In the space provided, write the first part of an abstract by briefly introducing the background, objective and methodology of your project.

Setting of work proposal 工作提案建構:

_____

_____

_____

Work problem 工作問題:

_____

_____

_____

Work objective 工作目標:

_____

_____

_____

Methodology to achieve objective　達成目標的方法：

_____

_____

_____

**G** Consider the following examples of how to write the first part of an abstract by briefly introducing the background, objective and methodology of your project.

*Statistics* 統計相關分類

**Setting of work proposal** 工作提案建構 Engineers heavily emphasize applicability and accuracy when using a process capability index to evaluate how a process performs. **Work problem** 工作問題 However, using conventional process capability indices to evaluate a non-normal distribution process often leads to inaccurate results. **Work objective** 工作目標 Therefore, this work presents an appropriate process capability index to evaluate non-normal distribution processes. **Methodology to achieve objective** 達成目標的方法 Clement's method is adopted to adjust the conventional indices. The bootstrap method is then applied to reduce the estimation error. Next, different non-normal distribution processes are evaluated by simulation to demonstrate the effectiveness of the proposed index. Additionally, a series of procedures is developed for engineers without a statistical background.

*Civil Engineering* 土木相關分類

**Setting of work proposal** 工作提案建構 Demands on the water supply in Taiwan have skyrocketed owing to the island's increasing population and elevated living standards. **Work problem** 工作問題 Eventually, the current reservoir supply will be insufficient if current personal and industrial demand persists. **Work objective** 工作目標 Therefore, this study describes a novel groundwater numerical model capable of analyzing the position and the quantities of optimal monitor wells in Choshuichi River Fan. **Methodology to achieve objective** 達成目標的方法 Geo-hydrological layers of existing and newly set wells are analyzed and a numerical model is developed. Setting parameters are then estimated from the covariance of regional variables. Additionally, a 3D groundwater numerical simulation is used to verify the geo-hydrological analysis and confirm the aquifer's position. Moreover, with the geo-statistical method, the layered structure is analyzed by the results of the groundwater numerical model. The statistical optimal monitoring design is also determined by variance reduction analysis.

*Civil Engineering* 土木相關分類

**Setting of work proposal** 工作提案建構 Although the reservoir supply in southern Taiwan can satisfy the current demand for water, **Work problem** 工作問題 increasing demand will soon surpass the system's capacity. **Work objective** 工作目標 Therefore, this study presents an effective water distribution optimization model capable not only of accurately reflecting how multi-objectives compete with each other, but also of estimating the available releases of multi-reservoirs. **Methodology to achieve objective** 達成目標的方法 A simulation model is developed to distribute the releases among reservoirs based on the operating rules of multi-reservoirs. These releases are then used to evaluate the agricultural and non-agricultural objectives. Next, the non-inferior solution set which reflects the impact on the relationship between agricultural and non-agricultural objectives is estimated by a multi-objective genetic algorithm.

*Industrial Engineering* 工業工程相關分類

**Setting of work proposal** 工作提案建構 An increasing number of wafer fabs use a control chart to detect assignable causes, making it extremely difficult for engineers to control effectively the wafer process. **Work problem** 工作問題 However, conventional control charts are designed for manufacturing systems with only one source of variation, making it impossible to control several independent sources of variation and often leading to ineffective and futile searches for assignable causes. **work objective** 工作目標 For use in wafer fabs, this work presents a competent on-line control process capable of detecting assignable causes concealed behind multiple characteristics and multiple readings in a manufacturing system with several sources of variation. **Methodology to achieve objective** 達成目標的方法 Principal component analysis (PCA) is performed to form new variables, which are the key components of original multiple characteristics in a manufacturing system. Their formation decreases the number of control charts since PCA reduces the number of related features. A multivariate exponential weight moving average (MEWMA) control chart is then used to verify whether the process is controlled. Additionally, for a situation in which MEWMA indicates that the process is out of control, three unique EWMA control charts of different sources of variation are used to identify the source of inaccurate variation.

*Statistics* 統計相關分類

**Setting of work proposal** 工作提案建構 Decision-making environments are increasingly complex, **Work problem**工作問題 accounting for why conventional evaluation models are concerned only with economic factors and neglect those factors that can not be evaluated in money. Such negligence may lead not only to an incomplete evaluation, but may also cause decision makers to select an inappropriate scheme. **Work objective** 工作目標 Therefore, this study presents a novel evaluation model capable of selecting natural gas bus brands. **Methodology to achieve objective** 達成目標的方法 Natural gas bus brands are selected by the model by performing cost effectiveness analysis. Each criterion is also evaluated for cost and effectiveness. Two methodologies of multiple attribute decision making (MADM), including the technique for order preference by similarity to ideal solution (TOPSIS) and the analytic hierarchy process (AHP), are then used to rank all viable alternatives to bus systems from a complete perspective.

*Statistics* 統計相關分類

**Setting of work proposal** 工作提案建構 Despite their extensive use in solving problems related to ordered categorical data quality, **Work problem** 工作問題 conventional scored methods are unnecessarily complex and inaccurate in estimating the dispersion effect. **Work objective** 工作目標 Therefore, this work develops an efficient response surface method capable of optimizing ordered categorical data process parameters since setting the process parameters leads to optimization of the location and dispersion effects. **Methodology to achieve objective** 達成目標的方法 This novel method with simplified calculations of the mean and standard deviation of ordered categorical responses is used to estimate the location and dispersion effects. Additionally, regression models are used to relate the location and dispersion effects to the controlled factor levels. An optimal combination of process parameters is also obtained using the dual response surface methodology.

## *Industrial Engineering* 工業工程相關分類

**Setting of work proposal** 工作提案建構 Although process quality and delivery time have been increasingly emphasized by industry, **Work problem** 工作問題 conventional process capability indices (PCIs) can neither objectively assess quality and delivery time nor identify the relationship between PCIs and yield rate. **Work objective** 工作目標 Therefore, this work presents an efficient hypothesis testing procedure for PCIs, capable of assessing the operational cycle time (OCT) and delivery time (DT) for VLSI. **Methodology to achieve objective** 達成目標的方法 The PCIs of OCT and DT are defined, and the UMVU (uniformly minimum variance) estimators of the studied PCIs are derived under the assumption of a normal distribution. These estimators are then used to construct the one-to-one relationship between the PCIs and the conforming rate of DT (or OCT). Finally, a hypothesis testing procedure for PCIs is developed.

## *Distance learning* 遠距教學相關分類

**Setting of work proposal** 工作提案建構 Although an increasing number of Internet-based learning environments are available to enhance knowledge construction, **Work problem** 工作問題 the learning activities of these learning environments over rely on in-group cooperation,

*Unit*
*Five*

Writing the abstract (part one): briefly introducing the
background, objective and methodology
摘要撰寫（第一部分）：簡介背景、目標及方法

erroneously implying that group members participate equally. **Work objective** 工作目標 Therefore, this work presents an Internet-based constructive learning environment that allows participants interactively to link their concept maps for accumulative learning. **Methodology to achieve objective** 達成目標的方法 Three groups (each including 10 to 11 students) were formed, in which a chapter was assigned to each group for reading and concept mapping. Three groups (each comprised of 10 to 11 students) were formed, and a chapter assigned to each group for reading and concept mapping. Instructed to construct individual concept maps in a knowledge construction environment, each student could only view his or her own concept map to prevent students from copying their peers. Next, after reading concept maps constructed by the other groups, students selected the best concept map. Students were then instructed to link their concept maps to the best one. Finally, the concept maps of each student were evaluated by voting

*Computer Science* 資訊科學相關分類

**Setting of work proposal** 工作提案建構 Although an increasing number of mathematical courses are available on the Web, **Work problem** 工作問題 plain text as their user interface have difficulty in editing complex mathematical equations. Consequently, users of conventional editors spend an excess of time to express equations that include more than ten mathematical symbols and students spend longer to learn how to use the teaching system. **Work objective** 工作目標 Therefore, this work presents a user-friendly editor of mathematical symbols capable of using a graphic user interface to reduce the learning time and provide an easier way to edit complex mathematical equations on the web.

*Distance Learning* 遠距教學相關分類

**Setting of work proposal** 工作提案建構 Despite their increasing availability on the Internet, **Work problem** 工作問題 distance learning courses lack feasible strategies for assessing student performance, ultimately inhibiting distance learning. **Work objective** 工作目標 Therefore, this work presents a networked peer assessment system capable of supporting instruction and learning to analyze students'learning outcomes in higher education. **Methodology to achieve objective** 達成目標的方法 For classroom use in distance learning courses, students can review the homework of their peers and receive comments. At the end of a semester, educators can easily access the student profiles easily via this system for further analysis.

*Computer Science* 資訊科學相關分類

**Setting of work proposal** 工作提案建構 Although extensively used in nonlinear programming, **Work problem** 工作問題 piecewise linearization algorithms require too much time to obtain an optimum solution. **Work objective** 工作目標　Therefore, this study presents an enhanced piecewise

215

linearization algorithm, capable of obtaining the global optimum of a nonlinear model, for use in a web based optimization system. **Methodology to achieve objective** 達成目標的方法 A web-based optimization system is implemented based on the enhanced algorithm and using a dynamic linking library procedure. The system is then linked to many other mathematical methods, for example, LINGO, to solve a nonlinear problem by integrating concurrent methods. Next, user specified problems are stored in a database storage system. Additionally, the solution is derived to guarantee the global optimum with an acceptable error rate.

*Information Management* 資訊管理相關分類

**Setting of work proposal** 工作提案建構 Although 3D models are extensively adopted in multimedia applications owing to their relatively low cost and ease with which they can construct animated 3D objects **Work problem** 工作問題 conventional 3D models are too time consuming and inaccurate when constructing digital objects since they manually retrieve 2D images. **Work objective** 工作目標 Therefore, this work presents an efficient face model capable of formulating the 3D image of an individual's face from three 2D images. **Methodology to achieve objective** 達成目標的方法 Three 2D images are obtained simultaneously with three general cameras. The images are then transformed to a personal computer. Next, the proposed model is used to formulate digitally the 3D face.

*Information Management* 資訊管理相關分類

**Setting of work proposal** 工作提案建構 Despite the increasing use of geographic-based information in daily living, **Work problem** 工作問題 the passive mode of accessing information fails to transmit effectively geographic-based information to PDA users. **Work objective** 工作目標 Therefore, this work presents a novel GIS-based architecture that supports an automatic reporting service through handheld mobile devices. **Methodology to achieve objective** 達成目標的方法 A GPS module is used to obtain the coordinates of the PDA user. The coordinates are then transmitted to the back-end server through a wireless network and used as the filter to query the database. Next, query results are sent back to the PDA, triggering an event to inform the PDA user.

*Information Management* 資訊管理相關分類

**Setting of work proposal** 工作提案建構 Intranets are extensively used by enterprises to accelerate commercial activities. **Work problem** 工作問題 However, conventional network management systems are too expensive and complicated to be implemented in an enterprise's Intranet. **Work objective** 工作目標 Therefore, this work presents a novel web-based system in a PC-LAN environment capable of identifying network problems. **Methodology to achieve objective** 達成目標的方法 A graphic interface is designed to illustrate the network structure of an enterprise. A free software

*Unit*
*Five*

Writing the abstract (part one): briefly introducing the
background, objective and methodology
摘要撰寫（第一部分）：簡介背景、目標及方法

program known as Multi Router Traffic Grapher (MRTG) is then applied to communicate with network devices. Return values from network devices are then filtered to detect potential network problems. Additionally, the visualized information is updated on the user interface of a web page.

*Mechanical Engineering* 機械相關分類

**Setting of work proposal** 工作提案建構 Although affecting the reliability of flip chip packaging under thermodynamic loading, **Work problem**工作問題 geometric parameters of a C4 type solder joint can not accurately predict. **Work objective** 工作目標 Therefore, this study presents an analytical geometry method capable of accurately predicting the geometric parameters of a C4 type solder joint in flip chip technology after a reflow process. **Methodology to achieve objective** 達成目標的方法 The differential equation of curvature is derived from the well-known Laplace-Young equation. The contact angle between the solder and substrate and the surface tension of the solder is then measured. The free surface energy of the substrate and the interfacial tension between the solder and the substrate is also obtained from Young's equation. Next, the differential equation of curvature is numerically solved to obtain the geometric parameters.. Finally, the geometric data of the solder joint obtained herein is used as an input in the finite element model to analyze the stress/strain distribution, thermal fatigue life and reliability of the electronic packages.

*Computer Science* 資訊科學相關分類

**Setting of work proposal** 工作提案建構 Although many machine learning and optimization application-related problems are solved by GAs with the multi-state property, **Work problem** 工作問題 conventional methods cannot solve multi-state problems. **Work objective** 工作目標 Therefore, this work presents a novel GA model capable of deriving an optimal solution for each state in a multi-state problem. **Methodology to achieve objective** 達成目標的方法 The proposed fuzzy polyploidy, a multi-state chromosome coding scheme, is used to describe the solution of a multi-state problem. Moreover, an adaptive genetic structure model is adopted to derive an appropriate polyploidy structure for practical applications. The proposed model consists of three structural level operations, including structural expansion, structural deletion, and structural coercion, to simulate the natural random variation.

*Physics* 物理相關分類

**Setting of work proposal** 工作提案建構 Weak disorder leads to two critical behavior scenarios of the $O(m)$ model, **Work problem** 工作問題 thus making such behavior of materials described by the weakly diluted $O(m)$ model unclear with respect to the dimensions of the order parameter. **Work objective** 工作目標 Therefore, this study presents a novel estimation method capable of obtaining the marginal number of order parameter

*Unit*
*Five*
Writing the abstract (part one): briefly introducing the
background, objective and methodology
摘要撰寫（第一部分）：簡介背景、目標及方法

dimensions. **Methodology to achieve objective** 達成目標的方法 Perturbation theory is expanded for mc. The resummation procedure is then applied to those expansions. Next, resummation results are analyzed to estimate the marginal dimension of a weakly diluted O(m) model.

*Physics* 物理相關分類

**Setting of work proposal** 工作提案建構 While ionizing irradiation stimulates material damage, the radiation-stimulated ordering effect (RSOE) is found in various materials. For instance, low-dose electron and X-ray irradiation enhance parameters in III-V and Si. **Work problem** 工作問題 However, the RSOE mechanisms in II-VI semiconductors remain unknown, making it impossible to implement process applications capable of increasing the efficiency of barrier structures. **Work objective** 工作目標 Therefore, this work investigates RSOE in II-VI semiconductors and constructs a related model. **Methodology to achieve objective** 達成目標的方法 Voltage-current characteristics (VIC), voltage-capacity characteristics (VCC) and capacity-modulated spectra for barrier structures are measured at various irradiation doses, providing the preliminary experimental data. Based on that data, the control parameters are derived. Following Hall experiments to consider bulk effects, the parametric changes are analyzed and synthesized.

*Physics* 物理相關分類

**Setting of work proposal** 工作提案建構 The surface control method based on exoelectron emission (EEE) is a precise and non-destructive relaxational method which identifies and predicts the early stages of material destruction. Although the EEE from alkali halide crystals has been extensively studied theoretically and experimentally, **Work problem** 工作問題 the theoretical description of EEE from alkali halide crystals remains insufficient, causing difficulties when analyzing and interpreting exoemission current-related data. **Work objective** 工作目標 Therefore, this study attempts to calculate the theoretical exoelectronic energy spectra to determine whether the defect-recombination mechanism is responsible for EEE from CsBr. **Methodology to achieve objective** 達成目標的方法 Crucial assumptions are presented. The wave functions of exoelectrons in the initial and final states are then chosen. After the matrix element of the transition is derived, the parameters of the obtained energy spectra are calculated.

*Physics* 物理相關分類

**Setting of work proposal** 工作提案建構 Dielectric properties of DMAGaS-DMAAlS ferroelectrics produce variations in temperature and pressure behavior, **Work problem** 工作問題 accounting for why the physical nature of dielectric properties of DMAGaS-DMAAlS

*Unit*
*Five*
Writing the abstract (part one): briefly introducing the
background, objective and methodology
摘要撰寫（第一部分）：簡介背景、目標及方法

ferroelectrics remain unclear. **Work objective** 工作目標 Therefore, this study develops a novel order-disorder four-state model capable of theoretically describing the dielectric properties of the DMAGaS-DMAAlS ferroelectric crystals. **Methodology to achieve objective** 達成目標的方法 The model Hamiltonian is constructed to consider ordering processes in the subsystems of DMA groups. The interaction between groups in their various orientational states is then examined using the dipole-dipole approximation. Next, the thermodynamic characteristics of the model are calculated using the mean field approximation.

*Mechanical Engineering* 機械相關分類

**Setting of work proposal** 工作提案建構 Recently emerging semiconductor technologies have ushered in the feasibility of embedding several digital modules in a printed-circuit board (PCB) and combining a larger system with several sub-chips. Moreover, the system clock operates at a higher frequency than that of conventional systems. **Work problem** 工作問題 However, owing to different clock propagations, the sub-modules of a digital system may be asynchronous. **Work objective** 工作目標 Therefore, this work presents a SAR-controlled DLL deskew circuit capable of reducing the system clock skew problem. **Methodology to achieve objective** 達成目標的方法 SAR is used to control the DLL so that the deskew circuit can automatically optimize the delay time for clock synchronization. The SAR binary search method is then employed to reduce the lock time and maintain tight synchronization.　Moreover, with the clock-deskew buffers' using inverter chains, the deskew circuit can reduce the system clock skew and obtain a perfect output clock duty cycle. Furthermore, the delay buffer chain is adjusted so that the deskew circuit can fit in different operating environments.　Finally, the architecture is implemented using Synopsys and Cadence tools

## *Civil Engineering* 土木相關分類

**Setting of work proposal** 工作提案建構 Groundwater usage in Taiwan is increasing at an accelerated rate, **Work problem** 工作問題 subsequently leading to a growing incidence of groundwater pollution in major groundwater supply regions in Taiwan, such as the Ping-tung Plain. Contaminants that pollute the aquifer make drinking the water from that groundwater source impossible for several years. **Work objective** 工作目標 Therefore, this study presents a deterministic and stochastic model for simulating groundwater flow to assess monitoring network alternatives. **Methodology to achieve objective** 達成目標的方法 A conceptual model of groundwater flow is constructed for a particular site.　The conceptual model is then transformed into a deterministic numerical model using MODFLOW. Next, hydrogeological parameters are calibrated to increase accuracy by using the deterministic numerical model that incorporates MODFLOW. Additionally, a stochastic numerical model is developed by linking the Kalman filter with the system equation along with the observation equation of the deterministic numerical model.

*Electrical Engineering* 電子相關分類

**Setting of work proposal** 工作提案建構 Although carrier recovery can help a digital receiver to overcome offset-related problems, **Work problem** 工作問題 conventional methods cannot achieve wide range locking with fast acquisition and a better tracking performance just by a one-order loop filter in the PLL circuit.. **Work objective** 工作目標 Therefore, this work presents a numerical method to choose efficiently the bandwidth of the loop filter in the PLL circuit. An additional apparatus is also developed in the carrier recovery circuit to estimate offsets precisely. **Methodology to achieve objective** 達成目標的方法 A frequency detection apparatus is used in the carrier recovery circuit to lock the large frequency offset. The PLL circuit can then automatically switch the coefficients of the loop filter into distinct bandwidths to reduce vibration and to converge faster than conventional circuits.

*Electrical Engineering* 電子相關分類

**Setting of work proposal** 工作提案建構 Redeposition plays a major role in inhibiting HDP-CVD gap fill. Re-deposition of the SiOx molecule usually originates from two sources - Ar sputtering and molecule backscattering. **Work problem** 工作問題 In addition to causing the SiOx to be deposited on the opposite side, both sources easily close the trench and significantly reduce the gap-fill capacity. **Work objective** 工作目標 Therefore, this work presents a low re-deposition process in HDP CVD to increase the gap-filling capability of the next IC manufacturing generation. **Methodology to achieve objective** 達成目標的方法 The process pressure is decreased by reducing the deposited gas flow. After the gas sputtering is decreased, the RF power and gas ratio are adjusted accordingly.

*Electrical Engineering* 電子相關分類

**Setting of work proposal** 工作提案建構 Consumer demand for hard disc bandwidth has significantly increased in recent years, owing to the large size of files and disc size. **Work problem** 工作問題 However, conventional interfaces for hard discs cannot easily support such a high bandwidth and require many pins . **Work objective** 工作目標 Therefore, this work presents a novel connecting interface for hard discs that can be developed to support bandwidths of up to 150 MHz and minimize the number of pins to four. **Methodology to achieve objective** 達成目標的方法 This novel connecting interface with serial transmitting and receiving lines is used to facilitate communication between system and hard disc. The serial interfaces are then used to eliminate the cross talk between transmitted and received data, discovered in conventional interfaces, and easily support bandwidths of up to 150 MHz. Next, the minimal pin requirement is satisfied using the serial interfaces.

**Unit**
**Five** Writing the abstract (part one): briefly introducing the
background, objective and methodology
摘要撰寫（第一部分）：簡介背景、目標及方法

*Electrical Engineering* 電子相關分類

**Setting of work proposal** 工作提案建構 Despite the obstacles to accurately forecasting capacity, our company strives to enhance its market competitiveness by developing a more precise model. **Work problem** 工作問題 Although a viable solution to this problem, the dynamic capacity model requires a tremendous amount of data input. Generally, more data input implies a more accurate model. Restated, an accurate capacity forecast depends on sufficient input. According to our estimates, the dynamic capacity forecast and the actual throughput diverge by less than 10%. **Work objective** 工作目標 Therefore, this work presents a precise dynamic capacity model, capable of accurately forecasting capacity. **Methodology to achieve objective** 達成目標的方法 Parametric effectiveness analysis is performed to select accurate forecasting capacity from the input data. Each criterion is also evaluated under quantity and time phase categories, using historical data. Simulation is performed to create a mathematical environment to access the real world, where more data input implies a more accurate output. The parameters are also measured. Additionally, the wafer start schedule and equipment status (which are the primary parameters) are then regulated to rank all viable alternatives to forecast the capacity.

*Electrical Engineering* 電子相關分類

**Setting of work proposal** 工作提案建構 Product quality is essential in wafer production, especially in light of the shrinkage of electronic devices and complex circuits in a chipset. The test period is prolonged to ensure that most functional paths are detected before mass production is undertaken. **Work problem** 工作問題 However, the rising cost of tests have not improved overall testing ability. While numerous human resources and advanced test equipment have been used to develop a test program, functional failures have hardly diminished. **Work objective** 工作目標 Therefore, this work presents effective test parameters to demonstrate their utility in wafer testing. **Methodology to achieve objective** 達成目標的方法 Pertinent literature and relevant industrial specifications are surveyed. Engineering experiments are then performed to verify the hypothesis. Next, a monitor pin is added to acquire take measurements.

*Electrical Engineering* 電子相關分類

**Setting of work proposal** 工作提案建構 As IC designs become larger and increasingly complicated, hardware emulation is essential for their verification. **Work problem** 工作問題 However, emulators, such as our emulation solution, are too expensive and too slow. **Work objective** 工作目標 Therefore, this work presents an emulation method that costs less and performs more efficiently than conventional methods. **Methodology to**

*Unit*
*Five* Writing the abstract (part one): briefly introducing the
background, objective and methodology
摘要撰寫（第一部分）：簡介背景、目標及方法

**achieve objective** 達成目標的方法 Appropriate types of FPGA, for example, Virtex E of Xilinx or APEX of Altera, with a retail value of around US$10,000, are selected according to size and speed requirements, to emulate an actual chip. A FPGA synthesis tool, for example, FPGA compiler or Synplify PRO, is then adopted to map the design net-list into a edif file for FPGA. Next, relevant software for placing and routing FPGA is adopted to construct an emulation database in FPGA. Additionally, timing constraints are established to enhance the emulation. Moreover, the database is downloaded into FPGA to initiate emulation and debug new designs

*Finance* 財務相關分類

**Setting of work proposal** 工作提案建構 An increasing number of interest rate derivative products priced by the forward interest rate have highlighted the empirical challenges of fitting forward rate yield curves to current market data. **Work problem** 工作問題 However, conventional methods which focus only on fitting the yield curve and transforming the forward rate curve from the yield curve give an unreasonable forward rate and an extremely high or negative number, such as 150% and -60%, respectively. **Work objective** 工作目標 Therefore, this work presents a novel numerical model capable of directly deriving a reasonable forward rate curve in one explicit function to correlate well with market data. **Methodology to achieve objective** 達成目標的方法 When including observable points, the forward rate curve of the proposed model is expressed as a function of a specified form with a finite number of parameters. The maximum smoothness term is then found within this parametric family to derive the model.

*Computer Graphics* 電腦製圖相關分類

**Setting of work proposal** 工作提案建構 3D applications play an increasingly important role in daily living, especially in entertainment. **Work problem** 工作問題 However, improvements in software and hardware can not keep pace with consumer preferences, making it impossible to design a high performance graphic chip bounded with drivers. **Work objective** 工作目標 Therefore, this work presents a high quality 3D graphics driver that performs excellently. **Methodology to achieve objective** 達成目標的方法 A robust and concise architecture is developed to satisfy the quality and performance requirements of a 3D driver. The hardware acceleration capabilities of the ICD architecture developed by SGI and Microsoft are then adapted to meet industrial standards. Next, the driver is developed by incorporating user-friendly properties .

*Environmental Engineering* 環工相關分類

**Setting of work proposal** 工作提案建構 Owing to environmental concerns, NF3 gas is gradually replacing CxFy gas in the dielectric CVD clean process. **Work problem** 工作問題 However, the global shortage of

NF3 clean gas accounts for its costing significantly more than CxFy clean gas, especially for high impurity ( > 4N ) NF3 clean gas. **Work objective** 工作目標 Therefore, this study presents a low impurity ( 3N ) NF3 clean gas, capable of reducing costs in the DCVD clean process. **Methodology to achieve objective** 達成目標的方法 The clean test rate is used to evaluate the 3N NF3 gas cleaning efficiency. The dielectric film's contamination test is then performed by ICP-MS and VPD-TXRF.

# *Unit Six*

William    Information Management（資訊管理）

Tolerable error    min

Reducing formulation costs

Creating more realistic digital objects

Melody    Information Management（資訊管理）

PDA

Commercialization in PDA Software

## Writing the abstract (part two): summarizing the anticipated results of the project and its overall contribution to a particular field

摘要撰寫（第二部分）：歸納希望的結果及其對特定領域的貢獻

**Vocabulary and related expressions**　相關字詞

Simulation results　模擬成果
A 3D face　3D構面（3D臉面）
optimizing　最有效的進行
manually retrieving　手動獲得
tolerable errors　可容忍的錯誤
constructing　建構
a digital face　數位的面孔
enhancing　提高
multimedia or animation applications　多媒體或卡通的運用
formulation costs　數位影像表達的成本
digital objects　數位的物體
proposed architecture　計畫的建築物（基地台）
automatically page　自動地瀏覽
PDA users　PDA使用者
handheld mobile device　手拿移動式的設備（PDA）
access information　使用資訊
geographic position　地理位置
filter　過濾器
commercialization　商業化
novel learning environment　新的學習環境
Genetic Algorithms (GA)　基因演算法
hand coding　手寫程式

**A** Write down the key points of the situations on the preceding page while the instructor reads aloud the script on page 340.

**Situation 1**

_____

_____

_____

**Situation 2**

_____

_____

_____

**Situation 3**

_____

_____

_____

Writing the abstract (part two): summarizing the anticipated results of the project and its overall contribution to a particular field

摘要撰寫（第二部分）： 歸納希望的結果及其對特定領域的貢獻

**B** Based on the three situations in this unit, write three questions beginning with *How*, and answer them.

**Examples**

*How does William's proposed model reduce the time to construct a 3D face by 10%?*

*By optimizing the 3D model rather than manually retrieving 2D images*

*How does Sherry's novel learning environment simplify the process of learning genetic algorithms?*

*By eliminating the need for hand coding GA programs*

1. _____

_____

2. _____

_____

3. _____

_____


C   Based on the three situations in this unit, write three questions beginning with ***What***, and answer them.

## Examples

*What allows PDA users to access information according to their geographic position?*

*Melody's novel architecture*

*What are some of the merits of Sherry's novel learning environment?*

*It increases the number of exercises to be practiced and improves the learning of GAs.*

1. _____

_____

Writing the abstract (part two): summarizing the anticipated results of the project and its overall contribution to a particular field

摘要撰寫（第二部分）： 歸納希望的結果及其對特定領域的貢獻

2. _____

_____

3. _____

_____

D Based on the three situations in this unit, write three questions whose answers are *Yes* or *No*, and answer them.

**Examples**

*Can Sherry's novel learning environment reduce the time required for students to complete a GA assignment?*

*Yes*

*Does William's proposed model manually retrieve 2D images?*

*No*

1. _____

_____

2. _____

_____

3. _____

_____

## E　Write questions that match the answers provided.

**Examples**

*How much time do students using Sherry's novel learning environment need to complete a GA assignment?*

*One week*

*Which tolerable errors can William's proposed model minimize?*

*Those associated with constructing a digital face*

Writing the abstract (part two): summarizing the anticipated results of the project and its overall contribution to a particular field

摘要撰寫（第二部分）：歸納希望的結果及其對特定領域的貢獻

1. _____

_____

Information

2. _____

_____

PDA users

3. _____

_____

The need for hand coding GA programs

# Unit Six

Writing the abstract（part two）：summarizing the anticipated results of the project and its overall contribution to a particular field

摘要撰寫（第二部分）：歸納希望的結果及其對特定領域的貢獻

1 Main results of the project 希望的結果
2 Contribution to field 領域的貢獻

*Unit*
*Six*

Writing the abstract (part two): summarizing the anticipated results of the project and its overall contribution to a particular field
摘要撰寫（第二部分）： 歸納希望的結果及其對特定領域的貢獻

Writing the second part of the abstract involves the following parts.

**1. Main results of the project 希望的結果. What are the anticipated results of your project? 你希望達成的結果?**

Consider the following examples.

◎Experimental results demonstrate the feasibility of using the studied PCIs to investigate and evaluate the suppliers' DT and OCT, underlying the choice of suppliers.

◎Based on portfolio and peer assessment, the networked portfolio system provides an environment that combines the flexibility of a network with the storage capacity of a computer. This system also allows students to collect their learning recode (including their homework), interact with peers, and critically reflect.

◎Experimental results indicate that the proposed algorithm reduces the computational time required to solve a nonlinear programming model to 50% of that required by piecewise linearization algorithms.

◎Simulation results indicate that the proposed architecture can automatically page PDA users through their handheld mobile devices when desired information becomes available. Moreover, the novel architecture allows PDA users to access information with their geographic position's functioning as a filter.

◎Simulation results indicate that the novel web-based system in a PC-LAN environment detects 85% of all network problems in 5 minutes by reducing the complexity of operating a network management system.

◎Theoretical results indicate that adopting the collective variables method allows the non-universal critical characteristics of a hard-sphere square-well binary symmetrical mixture to be quantified.

◎Analytical results indicate that the proposed method can be used to estimate the marginal dimension of an order parameter, corresponding to experimental data found in pertinent literature.

◎Experimental results indicate that the proposed method reduces the electrical resistance of thin films by 10 % when the thickness of metal films is constant.

◎Theoretical results indicate that the proposed model clarifies dielectric phenomena during the transition and disappearance of the ferroelectric phase under hydrostatic pressure.

◎Simulation results indicate that the proposed circuit can synchronize the 100~133 MHz system clock within 21 clock cycles.

**2. Contribution to field 領域的貢獻. How do you expect the results of your project to contribute to either a particular field or an industrial sector? 你的提案對相關工作領域的貢獻?:**

Consider the following examples.

◎Results in this study provide a valuable reference for government when selecting brands of bus systems.

◎Results in this study also confirm the reliability and validity of the proposed system as a viable assessment strategy for distance learning.

◎Results in this study demonstrate that, in addition to obtaining the global optimum in general nonlinear programming models within a tolerable error, the proposed algorithm significantly increases computational efficiency by decreasing the use of 0-1 variables.

◎Our results further demonstrate that the architecture proposed herein is highly promising for commercialization in PDA software.

*Unit*
*Six*

Writing the abstract (part two): summarizing the anticipated
results of the project and its overall contribution to a particular field
摘要撰寫（第二部分）： 歸納希望的結果及其對特定領
域的貢獻

◎Moreover, in addition to helping teachers and researchers to assess the higher-order thinking of students, the proposed system also helps students to train their critical thinking and analytical skills. Furthermore, the networked electronic portfolio system and peer assessment can be used at all educational levels.

◎Moreover, the learning environment eliminates the need for hand coding GA programs, simplifying the process of learning genetic algorithms.

◎Moreover, the proposed model minimizes the tolerable errors associated with constructing a digital face, enhancing multimedia or animation applications by reducing formulation costs and creating more realistic digital objects.

◎Results in this study demonstrate that the proposed method can also be used to design geometric parameters of a C4 type solder joint, capable of enhancing the reliability of a flip chip package and reducing its stress concentration.

◎Results in this study demonstrate that the proposed model accurately describes systems with more than two equilibrium positions on a site, extending conventional order-disorder models to a wider class of materials.

**F** In the space below, write the second part of the abstract by summarizing the anticipated results of the project and its overall contribution to a particular field.

Main results of the project 希望的結果:

_____

_____

_____

_____

Contribution to field 領域的貢獻:

_____

_____

_____

_____

*Unit*
*Six*

Writing the abstract (part two): summarizing the anticipated
results of the project and its overall contribution to a particular field
摘要撰寫（第二部分）： 歸納希望的結果及其對特定領
域的貢獻

G Consider the following examples of how to write
the abstract by summarizing the anticipated results
of the project and its overall contribution to a
particular field

*Statistics* 統計相關分類

**Main results of the project** 希望的結果 Statistical results indicate that the
proposed index evaluates non-normal distribution processes efficiently.
**Contribution to field** 領域的貢獻 Moreover, engineers can easily adopt
the proposed index and related procedures when comparing processes or
selecting an alternative supplier.

*Civil Engineering* 土木相關分類

**Main results of the project** 希望的結果 Simulation results indicate that
the proposed model can accurately estimate the new sitting wells precisely.
Those estimates can be used not only to design efficiently a ground water
network, but also to enable decision makers to identify optimal positions.
**Contribution to field** 領域的貢獻 In addition to minimizing the tolerable
errors, the proposed model can reduce the number of monitoring wells.

*Civil Engineering* 土木相關分類

**Main results of the project** 希望的結果 Simulation results indicate that the proposed model can determine when to transfer irrigation water from the irrigation association to non-agricultural areas, based on the non-inferior solutions, to offset water shortages during the dry season.. **Contribution to field** 領域的貢獻 The proposed optimization model provides a valuable reference for governmental authorities when drawing up water resource-related management strategies.

*Civil Engineering* 土木相關分類

**Main results of the project** 希望的結果 Simulation results indicate that the proposed stochastic model provides further insight into the uncertainty of the estimation error by adopting different groundwater monitoring network alternatives. **Contribution to field** 領域的貢獻 Additionally, the proposed model minimizes the costs of constructing a monitoring network by assessing monitoring network alternatives. Results in this study can be used to construct a real-time groundwater flow model for supporting the conjunctive use of water resources by combining the stochastic model and real-time groundwater level measurements.

*Industrial Engineering* 工業工程相關分類

**Main results of the project** 希望的結果 Simulation results indicate that,

245

*Unit Six* Writing the abstract (part two): summarizing the anticipated results of the project and its overall contribution to a particular field

摘要撰寫（第二部分）： 歸納希望的結果及其對特定領域的貢獻

in addition to detecting small shifts in a manufacturing system, the proposed control process can identify which sources of variation or characteristics are out of control. **Contribution to field** 領域的貢獻 Results in this study can provide a valuable reference for engineers when attempting to assess quickly the conditions of a manufacturing system. By effectively responding to this information, engineers can promptly adjust the manufacturing system to enhance wafer quality.

*Statistics* 統計相關分類

**Main results of the project** 希望的結果 Simulation results indicate that the proposed model can efficiently evaluate the cost and effectiveness of all viable alternatives to bus systems. **Contribution to field** 領域的貢獻 Results in this study provide a valuable reference for government when selecting brands of bus systems.

*Statistics* 統計相關分類

**Main results of the project** 希望的結果 Experimental results indicate that the proposed method accurately estimates the location and dispersion effects of an optimization procedure by solving problems related to ordered categorical data quality. **Contribution to field** 領域的貢獻 Moreover, the proposed method provides an optimization procedure with simplified calculations of ordered categorical data for engineers.

*Industrial Engineering* 工業工程相關分類

**Main results of the project** 希望的結果 Experimental results demonstrate the feasibility of using the studied PCIs to investigate and evaluate the suppliers' DT and OCT, underlying the choice of suppliers. **Contribution to field** 領域的貢獻 Moreover, PCIs can be used in uniform standards to assess DT and OCT for VLSI.

*Distance Learning* 遠距教學相關分類

**Main results of the project** 希望的結果 Experimental results indicate that the learning environment proposed herein prevents unequal participation in group activities. **Contribution to field** 領域的貢獻 This work provides a flexible cooperative-competitive model for educators to use in classroom activities.

*Distance Learning* 遠距教學相關分類

**Main results of the project** 希望的結果 This study indicated a significant relationship between students' attitudes and their performance, and identified the appropriate reliability and validity coefficients (e.g. r=.75**) of networked peer assessment. **Contribution to field** 領域的貢獻 Analytical results demonstrate the effectiveness of networked peer assessment, as evidenced by the significant relationship between students' attitudes and their performance. Results in this study also confirm its reliability and validity as an assessment strategy for distance learning.

*Computer Science* 資訊科學相關分類

**Main results of the project** 希望的結果 Based on portfolio and peer assessment, the networked portfolio system provides an environment that combines the flexibility of a network with the storage capacity of a computer. This system also allows students to collect their learning recode (including their homework), interact with peers, and critically reflect. **Contribution to field** 領域的貢獻 Moreover, in addition to helping teachers and researchers to assess the higher-order thinking of students, the proposed system also helps students to train their critical thinking and analytical skills. Furthermore, the networked electronic portfolio system and peer assessment can be used at all educational levels.

*Computer Science* 資訊科學相關分類

**Main results of the project** 希望的結果 Experimental results indicate that the proposed algorithm reduces the computational time required to solve a nonlinear programming model to 50% of that required by piecewise linearization algorithms. **Contribution to field** 領域的貢獻 Results in this study demonstrate that, in addition to obtaining the global optimum in general nonlinear programming models within a tolerable error, the proposed algorithm significantly increases computational efficiency by decreasing the use of 0-1 variables.

*Information Management* 資訊管理相關分類

**Main results of the project** 希望的結果 Simulation results indicate that the novel web-based system in a PC-LAN environment detects 85% of all network problems in 5 minutes by reducing the complexity of operating a network management system. **Contribution to field** 領域的貢獻 Moreover, the proposed system can ensure flexible and inexpensive network management for enterprises, by combining free software from the Internet.

*Mechanical Engineering* 機械相關分類

**Main results of the project** 希望的結果 Simulation results indicate that the proposed method can predict geometric parameters of a C4 type solder joint to within 5% of those obtained by a specific method found in the literature. **Contribution to field** 領域的貢獻 Results in this study demonstrate that the proposed method can also be used to design geometric parameters of a C4 type solder joint, capable of enhancing the reliability of a flip chip package and reducing its stress concentration.

*Computer Science* 資訊科學相關分類

**Main results of the project** 希望的結果 Simulation results indicate that the proposed GA model increases the accuracy of the optimum solution derived for a particular multi-state problem. **Contribution to field** 領域的

貢獻 Moreover, the proposed model enhances conventional genetic algorithms by systematically solving multi-state problems through the use of the polyploidy concept.

*Physics* 物理相關分類

**Main results of the project** 希望的結果 Analytical results indicate that the proposed method can be used to estimate the marginal dimension of an order parameter, corresponding to experimental data found in pertinent literature. **Contribution to field** 領域的貢獻 Moreover, applying several theoretical renormalization group methods to estimate the marginal dimension, in addition to those available in the literature, yields accurate results for a weakly diluted O (m) model, providing a complete picture of its critical behavior.

*Physics* 物理相關分類

**Main results of the project** 希望的結果 Theoretical results indicate that adopting the collective variables method allows the non-universal critical characteristics of a hard-sphere square-well binary symmetrical mixture to be quantified. **Contribution to field** 領域的貢獻 Moreover, this method can be used to calculate how thermodynamic functions depend on microscopic parameters near binary symmetrical mixture critical points, clarifying how microscopic parameters influence macroscopic critical behavior

250

*Physics* 物理相關分類

**Main results of the project** 希望的結果 Experimental and theoretical results clarify low-dose radiation processes in solids and enhance the parameters of barrier structures based on II-VI semiconductors. **Contribution to field** 領域的貢獻 Furthermore, this work extends the range of objects where the RSOE is observed and distinguishes between II-VI semiconductors and other ones (Si, III-V semiconductors) with respect to the RSOE mechanism.

*Physics* 物理相關分類

**Main results of the project** 希望的結果 Theoretical results indicate that the proposed calculation procedure obtains the energy spectra of the exoelectrons from CsBr. Contribution to field **Contribution to field** 領域的貢獻 Moreover, such an approach clarifies the EEE mechanism for CsBr, shedding further light on the formation of exoelectrons .

*Physics* 物理相關分類

**Main results of the project** 希望的結果 Experimental results indicate that the proposed method reduces the electrical resistance of thin films by 10 % when the thickness of metal films is constant. **Contribution to field** 領域的貢獻 Additionally, the proposed method eliminates the structure of scattering centers, providing further insight into the structure of thin films

*Unit*
*Six*

Writing the abstract (part two): summarizing the anticipated results of the project and its overall contribution to a particular field

摘要撰寫（第二部分）： 歸納希望的結果及其對特定領域的貢獻

and scattering centers.

*Physics* 物理相關分類

**Main results of the project** 希望的結果 Theoretical results indicate that the proposed model clarifies dielectric phenomena during the transition and disappearance of the ferroelectric phase under hydrostatic pressure **Contribution to field** 領域的貢獻 Results in this study demonstrate that the proposed model accurately describes systems with more than two equilibrium positions on a site, extending conventional order-disorder models to a wider class of materials.

*Physics* 物理相關分類

**Main results of the project** 希望的結果 The proposed model adopts statistical methodologies to increase the availability of the IMD World Competitiveness Model by 20%. **Contribution to field** 領域的貢獻 Results in this study demonstrate that the proposed method can enhance the world competitiveness ranking of Taiwan.

*Mechanical Engineering* 機械相關分類

**Main results of the project** 希望的結果 Simulation results indicate that the proposed circuit can synchronize the 100~133 MHz system clock within 21 clock cycles. **Contribution to field** 領域的貢獻 Moreover, the proposed circuit eliminates the system malfunction caused by the clock skew problem.

*Electrical Engineering* 電子相關分類

**Main results of the project** 希望的結果 Simulation results indicate that the novel design can lock a wide range of offsets of more than 100 KHz in a short acquisition time with a symbol error rate of less than 0.01. **Contribution to field** 領域的貢獻 Moreover, the improved carrier recovery on a digital receiver can track better than conventional models, making telecommunication products more competitive.

*Electrical Engineering* 電子相關分類

**Main results of the project** 希望的結果 Experimental results indicate that the proposed process can achieve an average aspect ratio of 2.7 for 0.2 um metal spacing. **Contribution to field** 領域的貢獻 By significantly enhancing the gap filling capacity to satisfy 0.15 um process requirements, the HDP-CVD process proposed herein can provide a good solution for the metal interconnection process when shrinking the dice size.

*Electrical Engineering* 電子相關分類

**Main results of the project** 希望的結果 Experimental results indicate that the novel serial interface can facilitate robust communication between system and hard disc and easily support bandwidths of up to 150 MHz. Moreover, the proposed interface can allow chipset vendors to integrate more functions into one chip than they did before. **Contribution to field** 領域的貢獻 In addition to making a high performance hard disc commercially feasible, the proposed interface can be easily upgraded in the future.

*Electrical Engineering* 電子相關分類

**Main results of the project** 希望的結果 Experimental results indicate that ranking forecasting capacity provides a more objective outcome with weights of the Marketing Department, than conventional methods. Moreover, ranking forecasting capacity can allow the sales department to identify the influence of customers' orders on alternatives. **Contribution to field** 領域的貢獻 The dynamic capacity model can also evaluate the cost and effectiveness of all viable alternatives to our fabrication, in which the quantity of work in process influences our inventory costs.

*Electrical Engineering* 電子相關分類

**Main results of the project** 希望的結果 The proposed test parameters reveal that more exactly knowing the shortage test items known reduces test time and increases product quality. **Contribution to field** 領域的貢獻 The proposed test parameters can be used to establish a productive qualification flow, improving capacity and reliability.

*Electrical Engineering* 電子相關分類

**Main results of the project** 希望的結果 Simulation results indicate that the proposed emulation method can of the costs and time of developing a 10MHz frequency by 95%, whereas the conventional method can only be used at 1MHz. **Contribution to field** 領域的貢獻 The proposed method can reduce testing time, include new devices that cannot function at 1MHz, and allow our company to verify the design with numerous platforms.

*Finance* 財務相關分類

**Main results of the project** 希望的結果 Numerical results indicate that the proposed numerical method provides the smoothest possible forward rate curve which is consistent with the chosen functional form and correlates with all observed market data on the yield curve. **Contribution to field** 領域的貢獻 Moreover, the numerical model provides an innovative means of directly fitting the forward rate with maximum smoothness and precision, providing a valuable solution for pricing interest rate derivatives.

*Electrical Engineering* 電子相關分類

**Main results of the project** 希望的結果 Simulation results indicate that our product can compete with other commercially available products, generating large revenues. **Contribution to field** 領域的貢獻 Our results further demonstrate that the proposed driver can also ensure flexible use by the end user.

*Environmental Engineering* 環工相關分類

**Main results of the project** 希望的結果 According to the clean test rate, the cleaning efficiency of the 3N NF3 gas is comparable with that of the high impurity gas. According to the dielectric film's contamination test, the cleaning efficiency of the 3N NF3 gas is comparable with that of both clean gases. However, the end-point time of the cleaning process in both clean gases is the same. **Contribution to field** 領域的貢獻 Results in this study demonstrate that the low impurity ( 3N ) NF3 clean gas can reduce the costs of the DCVD cleaning process by more than 50%.

# *Unit Seven*

 William    Information Management(資訊管理)

Reducing formulation costs

Tolerable error    min

Creating more realistic digital objects

 Melody    Information Management(資訊管理)

PDA

Commercialization in PDA Software

## Writing the work proposal

工作提案撰寫

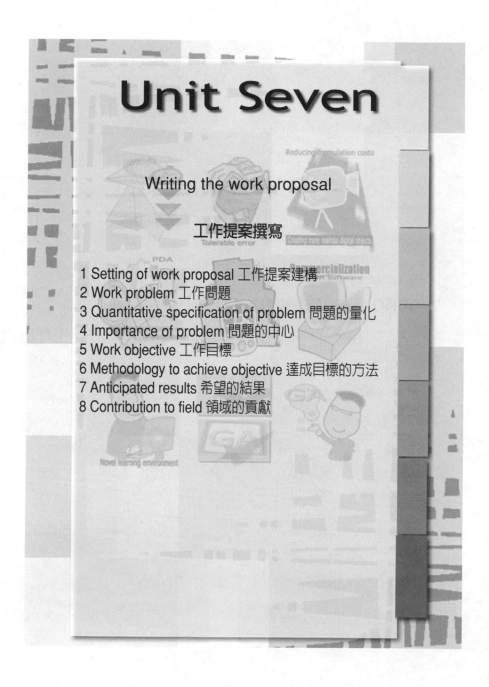

# Unit Seven

### Writing the work proposal

### 工作提案撰寫

1 Setting of work proposal 工作提案建構
2 Work problem 工作問題
3 Quantitative specification of problem 問題的量化
4 Importance of problem 問題的中心
5 Work objective 工作目標
6 Methodology to achieve objective 達成目標的方法
7 Anticipated results 希望的結果
8 Contribution to field 領域的貢獻

Write a work proposal of roughly 350-450 words.

Setting of work proposal 工作提案建構:

_____

_____

_____

_____

_____

_____

_____

_____

_____

Work problem 工作問題：

_____

_____

_____

_____

_____

_____

_____

Quantitative specification of problem 問題的量化：

_____

_____

_____

_____

_____

Importance of problem 問題的中心:

_____

_____

_____

_____

_____

Work objective 工作目標: Based on the above, we should

_____

_____

_____

**Methodology to achieve objective** 達成目標的方法:

_____

_____

_____

_____

_____

_____

_____

**Anticipated results** 希望的結果:

_____

_____

_____

_____

_____

_____

_____

_____

Contribution to field 領域的貢獻:

_____

_____

_____

Consider the following examples.

*Statistics* 統計相關分類

**Setting of work proposal** 工作提案建構 Decision-making environments are increasingly complex. For instance, selecting the criteria for developing a decision evaluation model is extremely difficult. While the inclusion of too few criteria in the evaluation model leads to incomplete results, too many makes the model too complex and difficult to evaluate. **Work problem** 工作問題 Conventional evaluation models are generally concerned only with economic factors and neglect factors that cannot be evaluated in money. However, large investment programs involve many intangible factors that cannot be valued, such as the satisfaction of related groups and potential environmental impact. In addition to an economic criterion, for example, when evaluating different brands of bus systems, many criteria are still applicable, such as the level of pollution from the bus and the drivability of the bus. **Importance of problem** 問題的中心 Adopting the conventional models is likely to cause decision makers to select an inappropriate scheme. For governmental investment decisions, implementing an inappropriate scheme will have large social costs and waste resources.

**Work objective** 工作目標 Based on the above, we should develop an efficient evaluationmodel capable of selecting natural gas bus brands. This model allows us to evaluate an appropriate number of criteria for cost and effectiveness. **Methodology to achieve objective** 達成目標的方法 To do so, natural gas bus brands can be selected by the model by cost effectiveness analysis. Two methodologies of multiple attribute decision making (MADM), including the technique for order preference by similarity to ideal solution (TOPSIS) and the analytic hierarchy process

(AHP), can then be used to rank all viable alternatives to bus systems from a complete perspective.

**Anticipated results** 希望的結果 As anticipated, the proposed model can evaluate exactly the relationship between the cost and effectiveness of all viable alternatives to bus systems. The evaluation results can provide not only economic information on all viable alternatives to bus systems, but many other comparable situations. **Contribution to field** 領域的貢獻 The proposed model can also provide a valuable reference for government when selecting brands of bus systems. The ranking methodology can provide a more objective outcome with weights of related decision groups, than can other methodologies. The ranking methodology can also provide a more flexible procedure with respect to the outcome's complexity. Moreover, the ranking outcome can allow decision makers to identify the order preferences for alternatives.

*Statistics* 統計相關分類

**Setting of work proposal** 工作提案建構 Despite their use in solving problems related to ordered categorical data, statistical methods are rarely applied to problems related to ordered categorical data quality. In addition to the accumulated analysis method (AA method) proposed by Taguchi, similar scored methods have also been proposed to solve problems of quality. Owing to its emphasis on the location effect, the AA method can accurately estimate only the location effect. To compensate for the limitations of the AA method, other scored methods have been proposed to accurately estimate the dispersion effect. **Work problem** 工作問題 Despite their extensive use in solving problems related to ordered categorical data quality, the AA method and the conventional scored methods are unnecessarily complex and inaccurate, confounding the location and dispersion effects. **Importance of problem** 問題的中心 Their implementation leads to an inaccurate optimal combination of process parameters, thus requiring much time and a higher cost in the product design stage .

**Work objective** 工作目標 Based on the above, we should develop an efficient response surface methodology capable of optimizing ordered categorical data process parameters, since setting the process parameters leads to optimization of the location and dispersion effects. **Methodology to achieve objective** 達成目標的方法 To do so, this novel method with simplified calculations of the mean and standard deviation of ordered categorical responses can be used to estimate the location and dispersion effects. Additionally, regression models can be used to relate the location and dispersion effects to the controlled factor levels. An optimal combination of process parameters can also be obtained using the dual response surface methodology.

**Anticipated results** 希望的結果 As anticipated, the proposed method with calculations of the mean and standard deviation of ordered categorical responses can accurately estimate the location and dispersion effects. In addition, it can be easily implemented and clearly separates the location effect from the dispersion effect. **Contribution to field** 領域的貢獻 The dual response surface methodology can be used to obtain an optimal combination of process parameters and help engineers set the controlled factors, and alleviate the quality problem related to ordered categorical data.

*Industrial Engineering* 工業工程相關分類

**Setting of work proposal** 工作提案建構 The build to order (BTO) model is gradually replacing the build to forecast (BTF) model. Process quality and delivery time have been increasingly emphasized by the highly competitive electronics industry. Additionally, many statisticians and engineers such as Kane, Kotz, and Pearn et al. have proposed process capability indices (PCIs) to assess the effectiveness of a process. **Work problem** 工作問題 Although performance has received considerable attention and many evaluation methods have been developed, manufacturing and delivery time have seldom been discussed. Additionally, conventional process capability indices (PCIs) can neither objectively assess quality and delivery time nor identify the relationship between PCIs and the conformation rate of DT and OCT. Utilizing point estimators of the

PCIs is generally an inaccurate means of estimating the real PCIs for suppliers. **Importance of problem** 問題的中心 Therefore, the lack of an effective performance index and an objective testing procedure will lead to inefficiency and a high overhead cost. Furthermore, if firms do not perform well in terms of quality and delivery, they will lose their market competitiveness. The outcome will also delay the firms' production.

**Work objective** 工作目標 Based on the above, we should construct an effective performance index (PCI) and develop an objective hypothesis testing procedure for PCIs, capable of assessing the operational cycle time (OCT) and delivery time (DT) for VLSI. A quality performance index is also used to assess operational cycle time and delivery time of VLSI. **Methodology to achieve objective** 達成目標的方法 To do so, the PCIs of OCT and DT can be defined and, then, the UMVU (uniformly minimum variance) estimators of the studied PCIs can be derived under the assumption of a normal distribution. Next, the above estimators can be used to construct the one-to-one relationship between the PCIs and the conforming rate of DT (or OCT). Finally, a hypothesis testing procedure for PCIs can be developed.

**Anticipated results** 希望的結果 As anticipated, this hypothesis testing procedure allows firms to assess the performance indices of the operation cycle time (OCT) and delivery time (DT) of VLSI, increasing the competitiveness of suppliers. Based on these performance indices, the corresponding tables of the excess time limit rate of OCT are also provided for manufacturing VLSI and DT, based on a supplier's schedule. These tables confirm the required performance index (PCI) value for manufacturers. Moreover, the hypothesis testing procedure for performance index is adopted to assess whether the OCT and DT satisfies the firm's requirements. **Contribution to field** 領域的貢獻 This work can provide a procedure for testing PCIs of OCT (DT) and a corresponding table of PCIs versus conforming rates of DT and OCT for suppliers. In addition to investigating the operational cycle time (OCT) of an individual manufacturing step for VLSI, the testing procedure can be used to assess the delivery time to satisfy customer requirements.

*Distance Learning* 遠距教學相關分類

**Setting of work proposal** 工作提案建構 An increasing number of educators provide  distance learning courses on the Internet (e.g. e.NCTU), as ushered in by the emergence of computer networks in the early 1980s to facilitate information sharing among researchers. Recent studies have considered the merits of using distance learning via a computer network to enhance instruction and learning. **Work problem** 工作問題 Despite their increasing availability on the Internet, distance learning courses lack feasible strategies for assessing student performance, ultimately inhibiting the effectiveness of distance learning. Despite their limitations, some web-based testing systems have been implemented to evaluate students' learning outcomes. For example, a student can answer questions on a test by studying prepared materials from a remote location. Notably, the drawbacks of a web-based testing system can be eliminated if a student's learning ability is evaluated through an activity. **Work objective** 工作目標 Based on the above, we should develop a networked peer assessment system capable of supporting instruction and learning to analyze students' learning outcomes in higher education. **Methodology to achieve objective** 達成目標的方法 For classroom use in distance learning courses, students can review the homework of their peers and receive comments. At the end of a semester, educators can access the student profiles easily via this system, for further analysis.

**Anticipated results** 希望的結果 As anticipated, the proposed networked peer assessment system can be used in distance learning courses, enabling students to review the homework of their peers and receive comments. At the end of a semester, educators can access the student profiles easily via this system, for further analysis. **Contribution to field** 領域的貢獻 Additionally, educators can identify the significant relationship between the students' attitudes and their performance, and identify the appropriate reliability and validity coefficients of networked peer assessment. While demonstrating the effectiveness of networked peer assessment, this study can also confirm its reliability and validity as an assessment strategy for distance learning.

*Computer Science* 資訊科學相關分類

**Setting of work proposal** 工作提案建構 An increasing number of mathematical courses are available on the Web. Today, the Internet is popular and the number of people who use the Internet is increasing. Many conventional courses are available on the web, including mathematics. Many researchers try to use networked learning technology to support students' learning. **Work problem** 工作問題 Conventional editors using plain text as a user interface have difficulty in editing complex mathematical equations. Mathematics' teaching materials include many mathematical symbols and equations. The mathematical symbols are graphical, and so mathematical equations are difficult to express in plain text. However, the user interfaces provided by conventional editors are plain text. **Quantitative specification of problem** 問題的量化 Users of conventional editors spend an excess of time to express equations that include more than ten mathematical symbols. **Importance of problem** 問題的中心 If users use plain text to express complex equations, they must be able to translate graphical mathematical symbols into plain text. Plain text is difficult to comprehend and most students never see mathematical equations in plain text and don't know how to use it.

**Work objective** 工作目標 Based on the above, we should develop a

mathematical editor using a Java applet on the web, capable of using a graphic user interface to edit mathematical symbols.

**Anticipated results** 希望的結果 As anticipated, a mathematical editor can be implemented to edit directly mathematical symbols with a graphic user interface. The proposed editor allows users to edit complex equations, even those containing more than 20 mathematical symbols. A graphic user interface is more intuitive than a text user interface. **Contribution to field** 領域的貢獻 The easy-to-use editor can be flexibly used on the Internet in mathematical courses, helping users more easily to edit complex mathematical symbols. The proposed editor can promote teaching mathematics on the web. It can help teachers to publish their mathematical material on the web and can be combined with other mathematical software (for example, Mathematica) to conduct mathematical examinations on the web.

*Distance learning* 遠距教學相關分類

**Setting of work proposal** 工作提案建構 Taiwan's Ministry of Education is increasingly emphasizing the use of multi-assessment in middle schools and universities. The means by which students' abilities are assessed heavily influences the development of a nation's labor force. A society without diverse individuals will eventually reduce a nation's competitiveness. Taiwan's Ministry of Education has stressed, in recent years, how this problem will influence the nation's future development, accounting for why it has executed a series of educational reforms. **Work problem** 工作問題　The conventional methods of assessing students' abilities fail to assess higher-order thinking owing to their inability to motivate students properly. **Importance of problem** 問題的中心　The failure will narrow the vision of students, and blot out students' learning autonomy. In conventional methods, students play a passive role and do not have any opportunity to exercise their higher-order thinking skills to control and reflect upon their learning outcomes.

**Work objective** 工作目標　Based on the above, we should develop a networked electronic portfolio system with peer assessment capabilities to provide a creative means of assessing the higher-order thinking of students. **Anticipated results** 希望的結果　As anticipated, the proposed networked portfolio system can observe the students' higher-level thinking by collecting the records of students' homework and interaction with peers on the networked electronic portfolio system with peer assessment. Educators can analyze the comments which students share with peers and what the content of students' reviews of peers' homework, to identify the higher-order thinking component. **Contribution to field** 領域的貢獻　This component can help educators to test and verify how the proposed system affects the higher-order thinking of students. Furthermore, the networked electronic portfolio system and peer assessment can be used at all

273

educational levels.

*Computer Science* 資訊科學相關分類

**Setting of work proposal** 工作提案建構 Piecewise linearization algorithms are extensively used in nonlinear programming. For instance, trading companies attempt to minimize the costs of factory-vendor transportation and ordering transactions. Such scenarios are normally formulated in a nonlinear format. Conventional algorithms can only obtain a local optimum in such scenarios. However, the difference between local and global optima leads to unexpected costs. **Work problem** 工作問題 However, piecewise linearization algorithms require too much time to obtain an optimum solution. For instance, while the objective function or constraint of a nonlinear problem is highly nonlinear, the solution and performance is always inadequate. Additionally, efficiency is more critical to the above problem than to other costs. Many engineers spend much to purchase equipment to solve their nonlinear problems in a relatively short time. **Quantitative specification of problem** 問題的量化 If piecewise linearization algorithms require more than 10 hours to obtain the optimal solution for general nonlienar programming problems, then equipment-related costs involved in obtaining the optimal solution are too high. **Importance of problem** 問題的中心 Related investigations can only formulate a smaller scale problem that represents only a small part of an actual situation. This outcome does not accurately reflect such a situation involves simulation and other related costs.

**Work objective** 工作目標 Based on the above, we should develop an enhanced piecewise linearization algorithm, capable of obtaining the global optimum of a nonlinear model, for use in a web based optimization system. **Methodology to achieve objective** 達成目標的方法 To do so, a web-based optimization system can be implemented based on the enhanced algorithm and using a dynamic linking library procedure. The system can then be linked to many other mathematical methods, for example, LINGO, to solve a nonlinear problem by integrating concurrent methods. Next, user specified problems can be stored in a database storage system. Additionally, the solution can be derived to guarantee the global optimum with an acceptable error rate.

**Anticipated results** 希望的結果 As anticipated, the enhanced piecewise linearization algorithm can reduce the computational time required to solve a nonlinear programming model to 50% of that required by piecewise linearization algorithms. Such an improvement not only significantly reduces computational time, but also allows users to make more efficient decisions. Moreover, the enhanced piecewise linearization algorithm can obtain the global optimum in general nonlinear programming models within a tolerable error and significantly increase computational efficiency by decreasing the use of 0-1 variables. **Contribution to field** 領域的貢獻 In addition to its usefulness in obtaining the optimum solutions in fields such as medicine, biology and science, the proposed algorithm can also provide the global optimum with a tolerable error. Furthermore, through the web-based optimization system proposed herein, user-specified problems can be stored in a database and used repeatedly. Via the proposed web-based system, the enhanced piecewise linearization algorithm can be applied in diverse fields such as medicine, biology and engineering. Through the user-friendly interface of the web-based system, users can easily and efficiently input their nonlinear model.

*Information Management* 資訊管理相關分類

**Setting of work proposal** 工作提案建構 3D models are extensively adopted in multimedia applications owing to their relatively low cost and the ease with which they construct animated 3D objects. Developing a concise and relatively easy means of constructing 3D faces has been heavily emphasized, particularly in computer animation. However, animating a human face is extremely difficult. Different models have been proposed to formulate 3D objects using expensive equipment and complex procedures. For instance, the Basin model uses a digitizer to retrieve 2D images and then transforms those images into 3D objects. **Work problem** 工作問題 Conventional 3D models are too time consuming and inaccurate when constructing digital objects since they manually retrieve 2D images. For instance, complex procedures involving the creation of digital objects require much time. Such approaches not only require too many steps in constructing individual faces, but also need expensive or precise instruments such as a 3D digitizer, laser scanner, or range finder. These obstacles create large overhead costs and inefficiency when formulating 3D faces, accounting for why many 3D facial formulation models are impractical for commercial use. **Quantititative specification of problem** 問題的量化 A situation in which conventional models require seven steps to construct a 3D model and their error rate exceeds 5% **Importance of problem** 問題的中心 makes it impossible to commercialize widely animation software and construct realistic digital objects. For example, although formulating 3D objects has received increasing interest for multimedia applications, conventional models can not meet commercial specifications.

**Work objective** 工作目標 Based on the above, we should develop an efficient face model capable of formulating the 3D image of an individual's face from three 2D images. **Methodology to achieve objective** 達成目標 的方法 To do so, three 2D images can be obtained simultaneously with three general cameras. The images can then be transferred to a personal computer. Next, the proposed model can be used to formulate digitally the 3D face.

**Anticipated results** 希望的結果 As anticipated, the proposed face model can reduce the time to construct a 3D face by 10% by optimizing the 3D model rather than manually retrieving 2D images. Additionally, the proposed model can precisely formulate an individual's face by using conventional peripheral equipment. Moreover, the constructed 3D face can not only be sculptured by automatic machinery, but can also simulate various expressions. The equipment required for constructing 3D digital objects includes three cameras, a personal computer, and common image software. Furthermore, steps to construct 3D digital objects in the proposed face model are simplified, reducing the related formulation costs and time to use the optimization methods. **Contribution to field** 領域的貢獻 The proposed face model can minimize the tolerable errors associated with constructing a digital face, enhancing multimedia or animation applications by reducing formulation costs and creating more realistic digital objects. The proposed model can be employed to digitize different real 3D objects.

*Information Management* 資訊管理相關分類

**Setting of work proposal** 工作提案建構 Geographic-based information is increasingly used in daily living. There is always the need to know, "Where am I going to?" and, "How do I get there?" in cases such as visiting a friend, going somewhere for business, sightseeing, making holiday plans, and so on. The required geographic information may be as simple an address, or as complicated as the complete path and estimation of time. Many kinds of geographical information are needed for various purposes. Many convenient applications that provide such information on PDAs are available in the market today. More and more people use these applications. **Work problem** 工作問題 However, the passive mode of accessing information fails to transmit effectively geographic-based information to PDA users. Users cannot use the system all the time, but some critical geographic-related information may appear when users are busy on other tasks. In some cases, they need to be informed immediately. If they have answers on time, they may save time, resolve an emergency, or even avoid an accident. But regretfully, no system capable of actively paging users exists . **Importance of problem** 問題的中心 The inability of PDA users to receive updated information in a timely manner will limit PDA use to within a narrow range. Such a limitation may discourage PDA use. Most people will not buy such an expensive "notepad" when they don't think its functions are useful. Only the evidently useful and convenient functions may change people's life styles and become essential. The conventional GIS systems in PDAs are not so advantageous as to be "killer applications".

**Work objective** 工作目標 Based on the above, we should design a GIS-based architecture that supports an automatic reporting service through handheld mobile devices. **Methodology to achieve objective** 達成目標的 方法 To do so, a GPS module can be used to obtain the coordinates of the PDA user. These coordinates can then be transmitted to the back-end server through a wireless network and used as the filter to query the database. Next, query results can be sent back to the PDA, triggering an event to inform the PDA user.

**Anticipated results** 希望的結果 As anticipated, the GIS-based architecture can automatically page PDA users through a wireless network when desired local information becomes available. Using the global positioning system (GPS) feature, the system will periodically send the position of the use to the server. The server will then check for any local news about disasters or roadblocks, and for advertisements. The results will be sent back to the user's machine and an event triggered to remind the user. **Contribution to field** 領域的貢獻 The GIS-based architecture can allow PDA users to access information with their geographic position's functioning as a filter. This avoids sending too much junk information to users. Limited by bandwidth and the speed of wireless network, this study also design strategies for data storage that involve a robust multi-tier architecture. The many geographic data are placed on the server, and only required data are transferred to the client host through the network.

*Information Management* 資訊管理相關分類

**Setting of work proposal** 工作提案建構 Intranets are extensively used by enterprises to accelerate commercial activities. Different operating systems and technologies are combined within the Intranet infrastructures of

enterprises. Therefore, maintaining a stable network environment is critical
. A network management system may help enterprises to maintain
efficiently a stable Intranet efficiently. **Work problem** 工作問題 However,
conventional network management systems are too expensive and
complicated for implementation in an enterprise's Intranet. **Quantitative
specification of problem** 問題的量化 Completely installing a network
management system requires at least five engineers' working full time for
more than 6 months. Moreover, enterprises must spend another six months
to train their employees and customize the system. Introducing a
conventional network management system to an enterprise also requires
additional machinery. Enterprises may need to modify the Intranet
infrastructure to fulfill the general requirements of a network management
system. **Importance of problem** 問題的中心 Additionally, enterprises
may have difficulty in upgrading software and hardware since they must
devote spend considerably to maintain the system, train operators, and
install the server.

**Work objective** 工作目標 Based on the above, we should develop a novel
web-based system in a PC-LAN environment capable of detecting network
problems. This system can provide an efficient and inexpensive solution for
an enterprise to maintain a stable network environment. **Methodology to
achieve objective** 達成目標的方法 To do so, a graphic interface can be
designed to illustrate the network structure of an enterprise. A free software

program known as Multi Router Traffic Grapher (MRTG), can then be applied to communicate with network devices. Return values from network devices can then be filtered to detect potential network problems. Additionally, the visualized information can be updated on the user interface of a web page.

**Anticipated results** 希望的結果 As anticipated, the proposed system can detect 85% of all network problems in 5 minutes by reducing the complexity of operating a network management system. An enterprise can smoothly apply the system without complex procedures. The proposed system can communicate with all remote network devices using the SNMP protocol. The system can automatically notify network administrators through e-mail, a pager, or voice mail. **Contribution to field** 領域的貢獻 The proposed system combines free software programs from the Internet to enhance the network management system. Combining a free software program Multi Router Traffic Grapher (MRTG) and a web-based integrated interface will greatly reduce developmental costs and efforts.

*Computer Science* 資訊科學相關分類

**Setting of work proposal** 工作提案建構 An increasing number of Genetic Algorithm (GA) courses are offered to solve optimization problems. As well known, GAs seek optimum solutions in a desired search space. Many GA-related workshops and conferences have been held over the past three decades, owing to their diverse industrial applications. Such courses instill in students the essential role of experimental design and simulation in learning GAs. **Work problem** 工作問題 However, students spend much time in coding programs for exercises when learning GAs, making it impossible for them to implement many GAs in a short time. Providing students with the opportunity to simulate GAs is essential in such a course. At the introductory level, teaching a new algorithm as an exercise is also expensive in terms of staff time, and can even be counterproductive since the student's written programs can introduce new problems. **Quantitative specification of problem** 問題的量化 Students who can implement only one GA in two weeks will learn GAs less effectively than those who can implement more Students cannot implement all the concepts introduced in a GA course, owing to time and course load constraints.

**Work objective** 工作目標 Based on the above, we should develop a novel learning environment capable of assisting students in flexibly learning genetic algorithms based on computer-assisted instruction. **Methodology to achieve objective** 達成目標的方法 To do so, several benchmark problems can be integrated in this environment. A mathematical expression can then be developed to provide users with fitness functions of GAs. Next, a case study involving a GA course can be presented to demonstrate the effectiveness of the proposed environment.

**Anticipated results** 希望的結果 As anticipated, the proposed environment can reduce the time required for students to complete a GA assignment to one week, increasing the number of practice exercises that can be implemented and allowing them to better learn GAs. During simulation, the learning environment immediately informs the students of the current operational status so that they can interact with the computer. Moreover, a complete evolutionary process can be logged in files for further analysis.

**Contribution to field** 領域的貢獻 The novel learning environment can eliminate the need for hand coding GA programs, simplifying the process of learning genetic algorithms. While recognizing the critical role that simulation plays in learning GAs, the proposed learning environment facilitates operation so that many common problems can be addressed. Moreover, this environment enables students to select desired system configurations, including structural settings and parametric selections, before simulation.

*Mechanical Engineering* 機械相關分類

**Setting of work proposal** 工作提案建構 As well known, geometric parameters significantly affect the reliability of flip chip packaging under thermodynamic loading. Based on the basic material mechanics, a higher solder joint height implies a larger shear force that the solder joint can bear. Moreover, several works have also concluded that cracks occur in a solder joint when a package is under critical thermodynamic loading. Furthermore, certain manufacturing issues, such as packaging falling-off problems, solder bridging and misalignment are also related to the design of the pad and solder joint. **Work problem** 工作問題   However, geometric parameters of a C4 type solder joint still can not accurately predict the reliability of flip chip packaging under thermodynamic loading.. Indeed, many engineers have tried to predict the geometric parameters of a C4 type solder joint using an energy-based method and analytical models. One of them simulated the C4 type solder joint as a semi-spherical high-lead bump buried into the eutectic solder; however the simulation lacks accuracy when the external loading increases. Besides, the above methodology ignores the effects of gravity. Moreover, the semi-spherical high-lead bump can not simulate the high-lead bump when the solder pad sizes vary. **Quantitative specification of problem** 問題的量化 If the geometric parameters can not reach an accuracy of 5%, engineers can neither predict the fatigue life nor enhance the yield of a flip chip package. **Importance of problem** 問題的中心 Under these circumstances, engineers can not predict the correct geometric parameters of a C4 type solder joint when the design factors vary. Therefore, engineers can neither predict the fatigue life nor enhance the yield of a flip chip package.

**Work objective** 工作目標 Based on the above, we should develop an analytical geometry method capable of accurately predicting the geometric parameters of practical C4 type solder joint in flip chip technology after a reflow process. **Methodology to achieve objective** 達成目標的方法 To do so, the differential equation of curvature is derived from the well-known Laplace-Young equation. Additionally, the contact angle between the solder and substrate and the surface tension of the solder can be measured. The free energy of the substrate and the interfacial tension between the solder and the substrate can then be obtained from Young's equation. Next, the differential equation of curvature is numerically solved to obtain the geometric parameters. Moreover, the geometric data of the solder joint can be used as an input in the finite element model to analyze the stress/strain distribution, thermal fatigue life and reliability of the electronic packages.

**Anticipated results** 希望的結果 As anticipated, the proposed methodology can calculate the effect of gravity on the buried high-lead solder bump instead of on the semi-spherical one. Furthermore, the analytical geometry method proposed herein can predict geometric parameters of a C4 type solder joint to within 5% of those obtained by a specific method found in the literature. **Contribution to field** 領域的貢獻 The proposed method can consider all of the design factors and can be used to investigate how those factors affect the final shape of a C4 type solder joint. Moreover, this study explores the effects of design factors on the reliability of flip chip technology. The proposed method can be used to design geometric parameters of a C4 type solder joint, capable of enhancing the reliability of a flip chip package and reducing its stress concentration. Additionally, the results of this study enhance the reliability and fatigue life of a flip chip package under thermodynamic loading.

*Computer Science* 資訊科學相關分類

**Setting of work proposal** 工作提案建構 Many machine learning and optimization application-related problems are solved by GA  with the multi-state property. For instance, in chess, a good player often employs various strategies based on his opponent's moves, the game's progress, or the chess clock. Therefore, an intelligent chess playing program should consider the multi-state property to perform more effectively.  Consider stock market investments as another example.  Investors adopt various strategies according to whether the market is up or down. This behavior implies that a decision support system for investment should consider the multi-state property. Applying various strategies to distinct states of a problem is natural. A different solution or strategy should be employed by varying problem solving states to achieve the global optimum. **Work problem** 工作問題 However, conventional methods cannot solve multi-state problems. Although genetic algorithms are widely applied to machine learning and optimization, conventional GAs have not received much attention . Conventional approaches can use human designed rule bases to represent a multi-state solution, and employ GAs to alter the rule base. However, to our knowledge, no systematic method has been developed for dealing with the multi-state property in a GA-implemented system.

**Quantitative specification of problem** 問題的量化 If the solution varies with the problem state in a multi-state problem, conventional methods neglect the multi-state property and yield an inaccurate and unfeasible solution. For example, an investment decision support system which adopts a single strategy, regardless of whether the market is up or down, leads to an unsuccessful investment. Although a rule base can be used to represent the multi-state property, designing the rule base requires many manhours.

**Work objective** 工作目標 Based on the above, we should design an effective GAs model capable of deriving an optimal solution for each state in a multi-state problem. **Methodology to achieve objective** 達成目標的方法 To do so, the proposed fuzzy polyploidy, a multi-state chromosome coding scheme, can be used to describe the solution of a multi-state problem. An adaptive genetic structure model can then be adopted to derive an appropriate polyploidy structure for practical applications. The proposed model consists of three structural level operations, including structural expansion, structural deletion, and structural coercion, to simulate the natural random variation.

**Anticipated results** 希望的結果 As anticipated, the proposed model can increase the accuracy of the optimum solution derived for a particular multi-state problem. Restated, a deeper multi-state property of a problem implies a more accurate solution achievable by the proposed model. For applications with only a signal state (without the multi-state property), the proposed model is comparable to conventional GAs and only requires a small amount of extra memory and computational time. **Contribution to field** 領域的貢獻 Moreover, the proposed GA method can enhance conventional genetic algorithms by systematically solving multi-state problems through the use of the polyploidy concept. Polyploidy encoding observed in nature provides a more flexible and dynamic encoding scheme than do conventional approaches. In practice, genetic algorithms have difficulty in obtaining optimum solutions when the chromosome structure is too complex because a complex structure always involves a large search space. The proposed simple to complex process, accompanied by a polyploidy model, facilitates the evolution of a complex structure that involves a large search space.

*Physics* 物理相關分類

**Setting of work proposal** 工作提案建構 As well known, weak structural disorder may significantly influence the critical behavior of different materials in particular magnets. It can not only alter the non-universal thermodynamic characteristics of a magnet but also change the universality class or modify the low-temperature phase behavior. Disordered magnets differ from each other according to the type of disorder. The weakly diluted $O(m)$ model describes random-site magnets. Its critical behavior may occur in two scenarios - with critical exponents of the pure $O(m)$ model and with changed ones.

The model universality class, determined by critical exponents, depends on the order parameter dimension, m. This dependence can be established by the Harris criterion, which states that disorder changes the universal critical properties of a pure model only if the heat capacity critical exponent, $\alpha$, of a pure model is positive. Within the hierarchy of the physical realizations of the O(m)- model, only the Ising model m=1 is characterised by $\alpha$ =0.109 $\pm$ 0.004 >0 , while the heat capacities of the xy- (m=2) and Heisenberg (m=3) models do not diverge at criticality: the corresponding critical negative $\alpha$ =-0.011 $\pm$ 0.004, $\alpha$ =-0.122 $\pm$ 0.009. Therefore, one can expect that only the weakly diluted quenched O(m=1) model belongs to a new universality class.

Indeed, experimental studies confirm the theoretical prediction. Much evidence collected in a recent review demonstrates novel critical behavior of magnetic systems described by the O(m)-model at m=1. The experimental value of the heat capacity critical exponent at the $\lambda$ -transition in He-4 corroborated that the system belongs to the xy-model universality class with no divergence of the heat capacity at the phase transition point. Subsequently, experiments on the critical behavior of He-4 in porous media confirmed the irrelevance of a weak quenched disorder in this case. For m=3, the experimentally measured critical exponents showed that the critical behavior of the weakly diluted Heisenberg model coincides with the error bars of the pure model. **Work problem** 工作問題 This lack of uniform data for various dimensions of an order parameter **Importance of problem** 問題的中心 makes it impossible to obtain the number of marginal dimensions that distinguishes the above scenarios.

**Work objective** 工作目標 Based on the above, we should develop an estimation method capable of obtaining the marginal number of order parameter dimensions. **Methodology to achieve objective** 達成目標的方法 To do so, perturbation theory can be expanded for mc. The resummation procedure can then be applied to those expansions. Next, resummation results can be analyzed to estimate the marginal dimension of a weakly diluted O(m) model.

**Anticipated results** 希望的結果 As anticipated, the proposed method allows us to estimate the marginal dimension of an order parameter, corresponding to experimental data found in pertinent literature. The obtained value of mc may predict some feature of critical behavior. **Contribution to field** 領域的貢獻 The proposed method also allows us to apply several theoretical renormalization group methods to estimate the marginal dimension, in addition to those available in the literature. This leads to two estimations for mc. Comparing those methods also allows us to select values which yield more accurate results for a weakly diluted O (m) model, providing a complete picture of its critical behavior.

*Physics* 物理相關分類

**Setting of work proposal** 工作提案建構 The interaction between binary mixture components leads to diverse phase behavior with respect to relative molecular sizes and the strengths of their interactions. Binary mixtures, in contrast to their constituent components, can exhibit three different types of two-phase equilibrium - vapor-liquid, liquid-liquid and gas-gas. The possibility of the realization of these phenomena and their priority depend both on the external conditions and the microscopic parameters of the mixture. Phase diagrams have been categorized. **Work problem** 工作問題 However, fully understanding the relationship between microscopic description and macroscopic phase behavior in a binary liquid system is extremely difficult. All classifications in previous literature were carried out phenomenologically or by mean field methods, preventing us from deriving from first principles even the simplest models of binary mixtures. **Importance of problem** 問題的中心 The failure to understand precisely which microscopic features form a particular phase topology makes it impossible to analyze the behavior of simple model thermodynamic functions near the critical point. Also unclear are the extent to which critical fluctuations affect the structure of the phase diagrams and whether the neglect of correlations by many analytical theories yields qualitatively and quantitatively incorrect phase diagrams.

**Work objective** 工作目標 Based on the above, we should elucidate the critical behavior of a binary symmetrical mixture using the collective variables method. **Anticipated results** 希望的結果 As anticipated, applying the collective variables method allows the non-universal critical characteristics of a hard-sphere square-well binary symmetrical mixture to be quantified. Free energy, entropy, heat capacity and equation of state, both above and below the critical point, can also be calculated. Using this approach, the phase diagram of the symmetrical mixture can be examined. **Contribution to field** 領域的貢獻 Importantly, the proposed investigation can calculate how thermodynamic functions depend on microscopic parameters near the binary symmetrical mixture critical points, clarifying how microscopic parameters influence macroscopic critical behavior.

*Physics* 物理相關分類

**Setting of work proposal** 工作提案建構As well known, ionizing irradiation causes material damage. The range/number of radiation effects in solids depends on both crystal structure and the type of ionizing irradiation. Additionally, the energy and dosage of irradiation essentially affect radiation processes. Ionizing irradiation excites both electronic and ionic subsystems of crystals. Relaxation processes occur after excitation, forming defects. Defects impair the solids, extensively stimulating material damage. However, the radiation-stimulated ordering effect (RSOE) is found in various materials. At least two competitive processes are observed in crystals and semiconductors under ionizing irradiation - 1) generation of radiation defects and 2) radiation-stimulated annihilation of defects. On a specific stage, the annihilation of irradiation defects can dominate the generation of defects, enhancing the structure. This stage is referred to as RSOE. Therefore, low-dose irradiation could serve as an effective method for increasing the reliability of semiconductors and the stability of parameters in the final stage of manufacturing. This fact substantially increases economical efficiency. **Work problem** 工作問題 Nevertheless, the RSOE mechanisms in II-VI semiconductors remain unknown, making it impossible to implement process applications capable of increasing the efficiency of barrier structures. **Importance of problem** 問題的中心 Although the application of RSOE could markedly increase product output generalization and analysis of RSOE remain an obstacle to further development.

**Work objective** 工作目標 Based on the above, we should investigate the RSOE in II-VI semiconductors and construct a related model. **Methodology to achieve objective** 達成目標的方法 To do so, voltage-current characteristics (VIC), voltage-capacity characteristics (VCC) and capacity-modulated spectra for barrier structures can be measured at various irradiation doses, providing the preliminary experimental data. Based on that data, the control parameters can then be derived. Following Hall experiments to consider bulk effects, the parametric changes can be analyzed and synthesized.

**Anticipated results** 希望的結果 As anticipated, low-dose radiation processes can be clarified in solids, improving the parameters of structures based on II-VI semiconductors. The proposed work can also elucidate low-dose radiation processes in II-VI semiconductors.. Thus, applying this model to barrier structure manufacturing allows us to predict accurately the irradiation conditions to enhance the parameters of structures based on II-VI semiconductors. **Contribution to field** 領域的貢獻 The range of objects in which the RSOE is observed can be expanded. II-VI and other (Si, III-V) semiconductors can also be compared in terms of the RSOE mechanisms. Analysis of RSOE in II-VI reveals the peculiarities of related objects. In contrast to III-V semiconductors, for which the properties of "pure" crystals are determined by impurities in II-VI semiconductors, the lattice "stoichiometric" defects prevails. Naturally, RSOE can be explained by the reconstruction of defect centers. Importantly, the proposed model considers the effect of the radiation-stimulated diffusion (RSD) of point defects.

*Physics* 物理相關分類

**Setting of work proposal** 工作提案建構 Dielectric properties of DMAGaS-DMAAlS ferroelectrics produce a variation in temperature and pressure behavior. A sequential change in paraelectric, ferroelectric and antiferroelectric phases was observed in the DMAGaS crystal. Ambient hydrostatic pressure narrows the temperature interval in which the ferroelectric phase exists, and, finally, suppresses it completely. However, for unknown reasons, the variety of phases of the isostructural DMAAlS crystal does not include the antiferroelectric phase. **Work problem** 工作問題 The physical nature of the dielectric properties of DMAGaS-DMAAlS ferroelectrics remains unclear. Although experiments prove reorientations of the DMA groups to be the origin of the observed phase transitions, an exhaustive description of the ordered state as a collective phenomenon should also consider interactions between those groups. Competition between interactions that tend to set up ferroelectric or antiferroelectric ordering, usually causes the phase transition of the type observed here. **Importance of problem** 問題的中心 Such a severe problem delays further experimental investigation and possible industrial application of these crystals. For instance, further insight into the ordering mechanism of the DMAGaS-DMAAlS ferroelectrics would allow one to better choose the aims of future experiments. For example, measuring the Y component of dielectric susceptibility enables the aforementioned competition of orderings be proven directly by observation of the possible point of the phase transition, "paraelectric-ferroelectric". Next, further investigation of reorientational hoppings of DMA groups could explain the difference between the behavior of DMAGaS and DMAAlS crystals.

**Work objective** 工作目標 Based on the above, we should develop a novel order-disorder four-state model capable of theoretically describing the dielectric properties of the DMAGaS-DMAAlS ferroelectric crystals. **Methodology to achieve objective** 達成目標的方法 To do so, the model Hamiltonian can be constructed to consider ordering processes in the subsystems of DMA groups. The interaction between groups in their various orientational states can then be examined using the dipole-dipole approximation. Next, the thermodynamic characteristics of the model can be calculated using the mean field approximation.

**Anticipated results** 希望的結果 As anticipated, the proposed model can clarify the dielectric phenomena during the transition and disappearance of the ferroelectric phase under hydrostatic pressure. Even the simplest phenomenological model based on the Landau expansion accurately predicts the aforementioned behavior of the crystals. However such an approach is limited to the vicinity of the phase transition point, giving a rather qualitative picture. A more accurate quantitative description can be achieved using a microscopic model and considering the symmetry of the system. **Contribution to field** 領域的貢獻 Moreover, the proposed model can accurately describe systems with more than two equilibrium positions on a site, extending conventional order-disorder models to a wider class of materials. Reorientational hoppings prohibit the proposed model from being expressed as a combination of Ising models. Our model is conceptually close to the Hubbard model (well known in the field of narrow band conductors), because it deals with a number of states and strong correlations. Besides the aforementioned ferroelectrics, the considered model can be used to describe KHCO3 (KDCO3) crystals and high-temperature YBa2Cu3O7-x superconductors.

*Information Management* 資訊管理相關分類

**Setting of work proposal** 工作提案建構 Taiwan's global competitiveness ranking in the IMD World Competitiveness Scoreboard is falling. Taiwan ranked 16th in 1998, 18th in 1999, and 22nd in 2000, with its lowest performance, 23rd, in the last decade of 1997. Although many countries originally ranked behind Taiwan, for example, Austria, Belgium, France, Iceland, Japan, New Zealand, and Sweden, they have since surpassed the island. Moreover, mainland China ranked 11 slots behind Taiwan in 1999, and 9 behind Taiwan in 2000. The gap between Taiwan and mainland China is shrinking. The narrowest gap over the last decade was of four places, in 1997. **Work problem** 工作問題 A contributing factor to the fall in Taiwan's IMD World Competitiveness ranking is the IMD Competitiveness Simulations' neglecting interactions among factors. These simulations involve twenty strengths and twenty weaknesses of each nation. If nations plan to upgrade their global competitiveness ranking, they need merely to improve their twenty weaknesses. Although the IMD runs the simulation to help policy makers to focus on and prioritize the key competitiveness issues facing their countries, they neglect the criteria's correlation. For example, if Taiwan improves its standing in relation to a weak criterion, Growth In Direct Investment Stocks Inward, its standing in relation to another strong criterion, Gross Domestic Savings Real Growth, will fall.

**Quantitative specification of problem** 問題的量化 More specifically, using the IMD Competitiveness Simulations possibly causes policy makers to reach inaccurate decisions. To gain entry into the WTO, Taiwan has drawn up a re-engineering plan, including financial liberalization and strengthening of financial supervision, to develop Taiwan as a global logistics center and a  regional operations center of the Asia-Pacific region\. Such a plan also aims to develop high-tech industries with venture capital in 2001. If the accuracy of the IMD Competitiveness Simulations cannot reach 50%, then increasing Taiwan's world competitiveness ranking is impossible.

**Work objective** 工作目標 Based on the above, we should develop a novel competitiveness model capable of integrating a global optimization algorithm into the IMD World Competitiveness Model. **Methodology to achieve objective** 達成目標的方法 To do so, populations can be collected from data of the IMD World Competitiveness Yearbook.Variables can then be analyzed.

**Anticipated results** 希望的結果 As anticipated, based on the IMD World Competitiveness Model, the proposed model can use cluster analysis to categorize countries into several classes. Each class can follow a unique approach to upgrade competitiveness ranking. This method can address interaction among factors. Additionally, the competitiveness model can increase the availability of the IMD World Competitiveness Model by 20%. A practical application of a known model can also be demonstrated.

**Contribution to field** 領域的貢獻 The model proposed herein can enhance Taiwan's world competitiveness ranking to within the top ten. Upgrading Taiwan's world competitiveness ranking is our government's mission. The IMD Competitiveness Simulations have some limitations. This work presents a novel means of eliminating such limitations. This work must provide policy makers with a clarified direction to enhance Taiwan's global competitiveness ranking.

*Mechanical Engineering* 機械相關分類

**Setting of work proposal** 工作提案建構 Recently emerging semiconductor technologies have ushered in the feasibility of embedding several digital modules in a printed-circuit board (PCB) and combining a larger system with several sub-chips. Moreover, the system clock operates at a higher frequency than do conventional system. Under these circumstances, the clock-skew problem becomes a critical issue. **Work problem** 工作問題 Owing to different clock propagations, the sub-modules of a digital system may be asynchronous. The process, voltage, temperature, and loading (PVTL) factors inevitably induce the clock-skew problem. Moreover, the skew problem will worsen as the clock'operational frequency increases, becoming a bottleneck in future-high-performance systems and possibly resulting in system malfunctioning. **Quantitative specification of problem** 問題的量化　A situation in which the system clock phase error caused by the clock-skew problem exceeds 5% leads to malfunctioning of the digital system, especially in a high performance system.

**Work objective** 工作目標 Based on the above, we should develop a SAR-controlled DLL deskew circuit, capable of reducing the system clock skew problem. **Methodology to achieve objective** 達成目標的方法　To do so, SAR can be used to control the DLL so that the deskew circuit can be automatically optimized for clock synchronization. The SAR binary search method is then employed to reduce the lock time and maintain tight synchronization. Moreover, with the clock-deskew buffers' using inverter chains, the deskew circuit can reduce the system clock skew and obtain a perfect output clock duty cycle. Furthermore, the delay buffer chain is adjusted so that the deskew circuit can fit in different operating environments. Finally, the architecture is implemented using Synopsys and Cadence tools **Anticipated results** 希望的結果 As anticipated, the proposed architecture can reduce the number of locking cycles in the

system clock more efficiently than can the conventional approaches.
**Contribution to field** 領域的貢獻 The scheme proposed herein to develop
an SAR-controlled DLL deskew circuit can reduce the system clock-skew
and achieve a perfect output clock duty cycle. Furthermore, the deskew
circuit can eliminate system malfunctioning caused by the clock skew
problem. Only the development of a deskew circuit can allow the system
clock to operate at Giga Hertz without interference.

*Civil Engineering* 土木相關分類

**Setting of work proposal** 工作提案建構 Groundwater usage in Taiwan is
increasing at an accelerated rate. Governmental authorities in Taiwan have
failed to effectively control groundwater usage owing to the inability to
construct an efficient pumping well management system that would detect
illegal pumping wells. Additionally, available groundwater information
such as hydraulic head and hydrogeology data is insufficient to detect
illegal structures.

**Work problem** 工作問題 Therefore, groundwater pollution is increasingly
common in major groundwater supply regions in Taiwan, such as in the
Ping-tung Plain. Excessive groundwater usage has led to stratum sag and
saline-water encroachment in aquifers since 1970 when the water table fell
below sea level in coastland areas of Taiwan. Additionally, the depth of
pumping wells has increased to supply fresh water for fish breeding

locations, ultimately leading to more pumping of groundwater in depth aquifer, increasing operational costs to do so, and adversely impacting the stratum sag. **Importance of problem** 問題的中心 Owing to the above trend, contaminants that pollute the aquifer have made drinking water impossible from that groundwater source for several years . Therefore, effectively monitoring the groundwater head and quality is of priority concern.

**Work objective** 工作目標 Based on the above, we should develop a deterministic and stochastic model for simulating groundwater flow to assess monitoring network alternatives. **Methodology to achieve objective** 達成目標的方法 To do so, a conceptual model of groundwater flow can be constructed for a particular site. The conceptual model can then be transformed into a deterministic numerical model by using MODFLOW. Next, hydrogeological paramters can be calibrated to increase accuracy by using the deterministic numerical model that incorporates MODFLOW. Additionally, a stochastic numerical model can be developed by linking the Kalman filter with the system equation along with the observation equation of the deterministic numerical model.

**Anticipated results** 希望的結果 As anticipated, the proposed stochastic model can provide further insight into the uncertainty of the estimation error of this model by adopting different groundwater monitoring network alternatives.　The proposed stochastic model provides the most detailed hydrogeological description of the Cho-Shui River Fan owing to its ability to incorporate hydrogeological data given by the monitoring wells for groundwater level and by geological logging. Based on the stochastic model and the deterministic numerical model, the Cho-Shui River Fan can be classified into four important aquifers and two regional aquitards.

**Contribution to field** 領域的貢獻 The proposed stochastic model can minimize the costs of constructing a monitoring network by assessing monitoring network　alternatives. Future investigations can construct a real-time groundwater flow model for conjunctive use in water resources by combining the stochastic model and real-time groundwater level measurements. In doing so, combining a real-time groundwater model with groundwater level monitoring wells allows governmental authorities　to detect the groundwater head drawdown　by　pumping wells.

*Statistics* 統計相關分類

**Setting of work proposal** 工作提案建構 Enterprises trade with each other by carefully considering business ratings to reduce investment risks. Each enterprise can use business ratings to achieve a balance between investment and over speculation. Varying risks display a significant geometric correlation with profits. When making an investment decision, a company should carefully assess their potential trade partners and the role of the investment scale. In doing so, they can select the best investment opportunity to maximize profits.

**Work problem** 工作問題 However, conventional mathematic models have difficulty in discriminating between multiple ranks. The neural network structure used in determining multiple ranks often results in an indefinite classification of business ratings. Additionally, the different levels of business rating can not be easily distinguished even when applying fuzzy theory to a neural network structure, thus making it impossible to distinguish between two multiple ranks. For instance, if two multiple ranks overlap, the business rating of some enterprises may be located in the center of those two ranks. **Quantitative specification of problem** 問題的量化 A prediction accuracy of only 60% among departed models exposes traders to unnecessarily high risks. **Importance of problem** 問題的中心 Over speculation will disrupt the financial flow of enterprises and easily corrupt their financial structure, ultimately leading to bankruptcy. Moreover, trading between enterprises resembles a connected network . Thus, bankruptcy of one firm will trigger a reaction throughout the financial sector.

**Work objective** 工作目標 Based on the above, we should develop a flexible and accurate neural network structure that applies artificial intelligence in fuzzy theory to business ratings and bankruptcy prediction. **Methodology to achieve objective** 達成目標的方法 To do so, the fuzzy theory can be incorporated to construct a neural network structure. Financial data of enterprises such as bankruptcy information, financial ratios, and other quality vectors can then be collected. Next, the credit ratings and bankruptcy information can be integrated to distinguish between an enterprise prone to bankruptcy and a fiscally sound one.

**Anticipated results** 希望的結果 As anticipated, the neural network structure that combines bankruptcy information and business ratings can increase the prediction accuracy of estimating a company's credit level to 90%. The proposed network structure can also clearly distinguish between a fiscally sound company and a fiscally unsound one owing to its ability to accurately predict factors that determine whether an enterprise is financially solvent. Based on our observation that industrially opposite ratios cannot replace industrial considerations, the neural network structure can incorporate the financial risks associated with an industrial sector in a company's annual report. Additionally, by using the upper and lower bounds of each financial ratio, the credit rating for different industrial sectors can be constructed to alleviate investment risks..

**Contribution to field** 領域的貢獻 Incorporating macroeconomic factors into the  neural network structure can increase its  predictiion accuracy, thus facilitating related investigations in assessing a mixture of investments. By adopting the neural network structure, investment planners can include bankruptcy prediction as a factor for evaluating a company's financial solvency. For instance, investment planners can combine the neural network structure and control charts such as CUSUM or EWMA  to monitor an enterprise's likelihood of becoming financially insolvent and its credit rating.  Therefore, more effective controlling an enterprise's financial rating ensures reliable business ratings in different industrial sectors. Creditors whom are capable of closely monitoring the fiscal performance of its trading partner on a quarterly or even monthly basis can reduce investment risk.  Moreover, three kinds of insolvent companies, i.e., short term, mid term, and long term ones, can be identified from their financial records. Making such a distinction allows creditors to estimate the potential insolvency of an enterprise as a reference for future investments.

*Civil Engineering* 土木相關分類

**Setting of work proposal** 工作提案建構 The increasing amount of organic solvents (NAPLs) used in industry has elevated the risk of leakage or spilling of organic compounds. When NAPLs are infiltrated through the saturated zone, some of the bulk liquid is trapped as immobile ones by capillary forces in a porous medium. In addition to having a low solubility in water, most organic solvents have difficulty in decomposing naturally. Therefore, the amount of residuals and transport mechanism of NAPLs in an aquifer must be determined.

**Work problem** 工作問題 Although numerous investigations have quantified residual non-aqueous phase liquid (NAPL) contamination in the subsurface, conventional methods such as core sampling and geophysical logging must be close to contaminant sources, thereby making it impossible to acquire sufficient samples and to identify where contaminants are distributed. Furthermore, improper use of those methods will yield unsatisfactory results and incur severe losses with respect to time and cost in groundwater remediation. **Quantitative specification of problem** 問題的量化 In addition, errors exceeding 20 % (as caused by uncertain and insufficient hydrology data) make it impossible to accurately predict their amounts. **Importance of problem** 問題的中心 An error rate exceeding 20% falls below R.O.C. environmental standards.

**Work objective** 工作目標 Based on the above, we should develop a non-aqueous phase liquid (NAPL) simulation model that contains several parameters acquired by experimental data. **Methodology to achieve objective** 達成目標的方法 To do so, the NAPL simulator can be estimated. The data set can then be obtained by a series of characteristic curve experiments that correspond to the simulation software. Next, the scaling rule theory can be applied to verify the effectiveness of the proposed NAPL simulation model when neglecting the hystersis effects.

**Anticipated results** 希望的結果 As anticipated, the proposed (NAPL) simulation model can predict the residual level of groundwater contaminants within an accuracy of 95% Scaling rule analysis can reveal that the experimental characteristic curve of the gas-water phases has a lower error than any other characteristic curves.

**Contribution to field** 領域的貢獻 Results in this study can provide a valuable reference for governmental authorities when attempting to achieve contaminants remediation. Additionally, the characteristic curve of the gas-water phases can be used in an on-site preliminary study to simulate the transport of NAPLs.

*Electrical Engineering* 電子相關分類

**Setting of work proposal** 工作提案建構 A digital receiver in a communications system must extract the correct symbols from data received by a tuner. The received data may become distorted, leading to an unacceptable error rate. Most distortion arises from multi-channel, timing offset, and carrier offset. Assume that the equalizer and the timing recovery apparatus can overcome the channel delay and mismatch of clock timing, respectively. How to resolve problems incurred from the carrier offset is thus of priority concern.

**Work problem** 工作問題 Despite a large carrier offset, wide range locking with a fast acquisition circuit design of carrier recovery helps a digital receiver to lock the carrier frequency in a relatively short time and with a tolerable error rate. However, conventional models with a digital phase-locked loop (PLL) circuit cannot do so since the loop filter unit inside the circuit is typically a one-order low-pass filter. Several methods have been applied to carrier recovery in order to enhance tracking performance. A notable resolution is Costas loop, which applies a frequency detector and a phase detector with their own loop filters. The outputs of the two loop filters are then combined to lock the precise carrier. Other approaches can be adopted to satisfy the various requirements of systems. For instance, a second or higher order loop filter is used in the PLL circuit to modify the conventional method. However, accurately representing the characteristics of a higher order loop filter may be difficult and impractical. Another modified circuit accesses a mechanism to switch the bandwidth of the one-order loop filter at a distinct time period, thus satisfying the tracking performance when the carrier offset is approximately ten times that of KHz. Nevertheless, as stated above, these methods can not lock the carrier frequency in a relatively short time and with a tolerable error rate.

**Quantitative specification of problem** 問題的量化 Under such circumstances, a loop filter with a wide bandwidth incurs a large vibration and ultimately a high error rate. Whereas a loop filter with a narrow one leads to slow convergence, even expending more than ten times the estimated acquisition time. Furthermore, a loop filter with a narrow bandwidth may not recover the carrier while the receiver suffers from a large frequency or phase offset. **Importance of problem** 問題的中心 Such behavior not only severely limits carrier recovery, but also ultimately leads to failure in the digital receiver.

**Work objective** 工作目標 Based on the above, we present a novel design to select the bandwidth of the loop filter in a PLL circuit efficiently. **Methodology to achieve objective** 達成目標的方法 To do so, a frequency detector following the phase detector can be adopted in the carrier recovery to lock the large frequency offset. While the carrier recovery circuit can receive the cross-talked symbols from the demodulation of the digital receiver, the phase detector can extract the decision-directed phase errors from these symbols. Such phase errors can then be fed into the frequency detector and the one-order loop filter, respectively. Next, computations can be performed to allow the frequency detector to transfer the phase errors into frequency data. In doing so, the transferred frequency information can accurately reflect the characteristics of the frequency offset. Moreover, the mechanism of the loop filter can switch the bandwidth at a distinct time period by a counter. Therefore, the coefficients of the bandwidth can be used in larger values to increase convergence. The coefficients can then be set as smaller ones to reduce vibration errors when the mechanism switches the bandwidth. Consequently, the joint activities of the phase detector, the frequency detector, and the loop filter can achieve wide range locking as well as a fast acquisition with an acceptable error rate.

**Anticipated results** 希望的結果 As anticipated, the novel design can lock a wide range of offsets more than 100 KHz in a short acquisition time with a symbol error rate less than 0.01. Therefore, the forward error correction (FEC) unit can decode the bit stream to become quasi error free. **Contribution to field** 領域的貢獻 Furthermore, the enhanced carrier recovery on a digital receiver can perform better in terms of tracking than conventional models, thus making communication products more competitive.

*Electrical Engineering* 電子相關分類

**Setting of work proposal** 工作提案建構 Consumer demand with respect to hard disc bandwidth has significantly increased in recent years owing to the large size of files, as well as elevated disc size. The storage device of a computer system's hard disc is we used as a secondary memory to boot up the system or start an application program. However, OS systems of Windows 2000 is significantly larger than that of Windows 3.1, necessitating an increase in the throughput when transferring data from the hard disc to the main memory, DRAM. Doing so would enhance performance and elevate the disc size to accommodate a large amount of data. **Work problem** 工作問題 However, the conventional interface for a computer system's hard disc cannot easily support such a high bandwidth and demands many pin counts to support such a requirement. Significant progress has been made in recent recent years to improve the parallel interface, with the throughput of a hard disc increased from 33 megabytes per second to 133 megabytes per second by adopting advanced technologies such as handshake protocols, clocking mechanism, and advanced fabrication. However, electromagnetic issues are of priority concern in light of efforts to make the throughput of a hard disc equal to or larger than 150 megabytes per second. In particular, emphasis must be placed on resolving the cross-talk interference induced in the connecting bus between the hard disc and the computer, **Quantitative specification of problem** 問題的量化 which can induce a failure rate exceeding 5% in manufacturing when the conventional interface must support 150 MHz bandwidth size. **Importance of problem** 問題的中心 This high failure rate makes commercialization impossible. Despite inserting a VDD or a GND between two signals to prohibit cross-talk interference between them, designers can still not eliminate such interference entirely when the throughput is increased to 150 megabytes per second or higher. Additionally, the failure rate in manufacturing will exceed 5 % because the cross-talk interference significantly degrades the data transmission, requiring that wafer manufacturers expend additional manpower to ensure that each product functions properly. Moreover, chipset providers can not integrate many functions into a chipset because the conventional interface requires many pins. Although an inexpensive package can accommodate limited pins, many pin counts are necessary to implement a multifunction chip. Consequently, the manufacturer must either increase both the pin count of the package and the production cost or minimize the pin count requirement for each function to integrate many functions in a chip.

**Work objective** 工作目標 Based on the above, we should develop a novel connecting interface for a hard disc to supporting up to 150 MHz bandwidth size and minimize the pin counts to 4 pins. **Methodology to achieve objective** 達成目標的方法 To do so, this novel connecting interface with serial transmitting line and receiving line can be used to facilitate communication between the system and a hard disc. The serial interfaces can then be used to eliminate the cross talk interference between transmitting data or receiving data found in conventional ones and easily support up to 150 MHz bandwidth size. Additionally, cross talk interference can be easily resolved because the serial interface transmits and receives data serially. Moreover, the minimal pin requirement can be satisfied using the serial interfaces, and minimal production's cost can be achieved as well.

**Anticipated results** 希望的結果 As anticipated, the novel serial interface can facilitate robust communication between system and hard disc and support up to 150 MHz bandwidths easily. Moreover, the proposed interface can allow chipset vendors to integrate more functions into one chip than they did before. **Contribution to field** 領域的貢獻 In addition to making a high performance hard disc commercially feasible, the proposed interface can be easily upgraded in the future.

*Electrical Engineering* 電子相關分類

**Setting of work proposal** 工作提案建構 High pin count packages are extensively adopted in high integrated electrical equipment. The increasingly MOS gate count requires a large die space, and a high volume signal I/O count also requires a larger die fan out environment. Unfortunately, an increasingly larger die size has decreased the die yield and increased the wafer cost. Therefore, wafer product designers continuously strive to achieve a balance between minimal die size and a feasible I/O fan out /layout feasibility. **Work problem** 工作問題 However, the conventional trial and error method to evaluate the maximum signal density so that a signal can successfully fan out requires too much time in obtaining an optimum solution. **Quantitative specification of problem** 問題的量化 For instance, when a designer predicts the die size and amount of signal I/O, more than 2 days are required to confirm whether this architecture can safely fan out on one time iteration. Additionally, nearly all of the cases are iterated more than three times. **Importance of problem** 問題的中心 The inability to construct a design database makes it impossible to shorten the IC package design schedule in the early stage. Therefore, IC package designers require a more efficient means of considering all factors to determine the minimum die size and feasibility of I/O fan out in order to construct a prototype model. Doing so would greatly facilitate a designer in shortening the IC design schedule and increasing design efficiency.

**Work objective** 工作目標 Based on the above, we should develop an efficient scheme capable of solving the improper die size estimation problem that simultaneously considers cost-related issues and design feasibility. **Methodology to achieve objective** 達成目標的方法 Using the proposed scheme, the optimal relationship between signal I/O count and die size can be derived. In doing so, the designer can reduce the die size and pay closer attention to electrical circuits inside the die. A database with an parameter can then be developed based on mathematical calculations as well as comparisons with previous database projects.

**Anticipated results** 希望的結果 As anticipated, the proposed scheme, in its mature stage of development, can facilitate IC design engineers in minimizing die size and achieving an appropriate package type, best package trace/ lead count fan out and excellent thermal / electrical performance as well as an optimized design. The proposed scheme can also reduce the schedule of trial-and-error by 30% over that of the previous stage, allowing us to obtain the minimal die size, lowest cost and optimum design. **Contribution to field** 領域的貢獻 The proposed scheme can shorten the design flow and dramatically reduce the time to market schedule. Owing to its excellent characteristics, the IC product can elevate our company's technical and market position.

# Answer Key

解　答

# Answer Key

Outlining the work proposal (part one): describing the project background
工作提案計畫（第一部分）：背景

## A.

### Situation 1

Many enterprises use Intranets to accelerate commercial activities However, conventional network management systems are too expensive and complicated to be implemented in a typical enterprise's Intranet. For instance, a network management system that costs more than 300,000 US Dollars to implement and requires additional machinery to operate makes upgrading hardware and software impossible.

### Situation 2

An increasing number of distance learning courses are available on the Internet. However, such courses lack feasible strategies to assess student performance. This lack of reliable assessment methods inhibits the effectiveness of distance learning.

### Situation 3

An increasing number of mathematics courses are available on the Web. However, conventional editors, using plain text as the user interface, causedifficulty in editing complex mathematical equations. For example, users of conventional editors spend excessive time in trying to express equations thatinclude more than ten mathematical symbols.

## B.

These are only possible questions.

What are increasingly available on the Internet?
Distance learning courses.
What do many distance learning courses on the Internet lack?
Feasible strategies to assess student performance.

What makes upgrading hardware and software impossible?
A network management system that costs more than 300,000 US Dollars to implement and requires additional machinery to operate

What do users of conventional editors spend excessive time in doing?
Trying to express equations that contain more than ten mathematical symbols.

## C.

These are only possible questions.

Which conventional editors cause difficulty in editing complex mathematical equations?
Those using plain text as the user interface.

Which methods are needed to improve the effectiveness of distance learning?
Reliable assessment methods.

Which systems are too expensive and complicated to be implemented in a typical
enterprise's Intranet?
Conventional network management systems.

## D.

These are only possible questions.

Why do users spend excessive time in trying to express equations that include more than ten mathematical symbols?
Because conventional editors, using plain text as the user interface, cause difficulty in editing complex mathematical equations.
Why is it impossible to implement conventional network management systems in a typical enterprise'sIntranet?
Because they are too expensive and complicated.

Why is it impossible to assess student performance on many distance learning courses
available on the Internet?
Because they lack feasible strategies to do so.

E.

Why is it impossible to implement conventional network management systems in a typical enterprise's Intranet?
Because they are too expensive and complicated.

Where are an increasing number of mathematics courses available?
On the Web.

Which editors cause difficulty in editing complex mathematical equations?
Conventional ones, using plain text as the user interface.

## A.
### Situation 1

Based on the above, Tom wants to develop an efficient evaluation model that can select brands of buses that run on natural gas. The proposed model can be used to evaluate the cost and effectiveness of all viable bus systems, as well as provide government with a valuable reference for selecting such systems.

### Situation 2

Based on the above, Susan wants to design a GIS-based architecture that supports a service that automatically reports to handheld mobile devices. The proposed architecture can automatically page PDA users through a wireless network when desired local information becomes available. Additionally, the proposed architecture allows PDA users to access information that is filtered according to geographic position.

### Situation 3

Based on the above, John wants to develop a networked peer assessment system capable of supporting instruction and learning to analyze the effectiveness of students' learning in higher education. The proposed system can identify a significant relationship between the students' attitudes and performance. The system can also obtain appropriate reliability and validity coefficients for networked peer assessment, demonstrating the feasibility of such assessment as an alternative strategy for distance learning.

## B.

These are only possible questions.

How can government use Tom's evaluation model?
As a valuable reference for selecting brands of buses that run on natural gas.

How does John plan to demonstrate the feasibility of his proposed system as an alternative strategy for distance learning?
By showing that his system can obtain appropriate reliability and validity coefficients for networked peer assessment.

How does John plan to analyze the effectiveness of students' learning in higher education?
By developing a networked peer assessment system capable of supporting instruction and learning.

## C.

These are only possible questions.

Why would government want to use Tom's evaluation model?
As a valuable reference for selecting brands of buses that run on natural gas.

Why does Tom want to develop an efficient evaluation model that can select brands of buses that run on natural gas?
To evaluate the cost and effectiveness of all viable bus systems.

## D.

These are only possible questions.
What can Tom's efficient evaluation model be used for?
To evaluate the cost and effectiveness of all viable bus systems, and provide government with a valuable reference for selecting such systems

What does Susan's GIS-based architecture allow PDA users to do?
Access information that is filtered according to geographic position

What does John believe is an alternative strategy for distance learning?
His networked peer assessment system.

E.

Who can be automatically paged through a wireless network when desired local information becomes available?
PDA users.

What does Susan want to design?
A GIS-based architecture that supports a service that automatically reports to handheld mobile devices.

What aspect of viable bus systems can Tom's model be used to evaluate?
Cost and effectiveness.

## A.
### Situation 1

Although process quality and delivery time have been increasingly emphasized by industry, conventional process capability indices (PCIs) can neither objectively assess quality and delivery time nor identify the relationship between PCIs and yield rate. Consequently, the lack of an effective performance index and an objective procedure leads to inefficiency and a high overhead cost. Furthermore, firms that perform poorly in terms of quality and delivery will lose their market competitiveness. Therefore, an efficient hypothesis testing procedure for PCIS must be developed, capable of assessing operational cycle time (OCT) and delivery time (DT) for VLSI.

### Situation 2

Despite the increasingly everyday use of geographical-based information, the passive mode of accessing information fails to transmit geographic-based information to PDA users effectively. Although perhaps unaware at the time, PDA users can benefit from such information, such as in cases of emergency. The inability of PDA users to receive updated information in a timely manner will severely limit the range of use PDAs. Such a limitation may discourage PDA use. Therefore, a GIS-based architecture that supports more interesting and useful functions than are supported by conventional architecture, must be developed.

### Situation 3

Taiwan's global competitiveness ranking in the IMD World Competitiveness Scoreboard is falling. Accordingly, the IMD Competitiveness Simulations neglect interactions among related factors. Use of the IMD Competitiveness Simulations can cause policy makers to make an inaccurate decision. Therefore, a novel competitiveness model must be developed, capable of integrating a global optimization algorithm into the IMD World Competitiveness Model.

## B.

These are only possible questions.

Why

Why does Jack feel that Taiwan's global competitiveness ranking in the IMD World Competitiveness Scoreboard is falling?
Because policy makers make inaccurate decisions when using IMD Competitiveness Simulations.

Why does Jerry believe that inefficiency and a high overhead cost occur in a company?
Because of a lack of an effective performance index .

Why would consumers be discouraged from using PDAs?
Because of the inability to receive updated information in a timely manner.

## C.

These are only possible questions.

How

How does Jack want to prevent policy makers from making inaccurate decisions when using the IMD Competitiveness Simulations?
By developing a novel competitiveness model capable of integrating a global optimization algorithm into the IMD World Competitiveness Model.

How might firms lose their market competitiveness?
By performing poorly in terms of quality and delivery.
How is Taiwan doing in the IMD World Competitiveness Scoreboard?
Its global competitiveness ranking is falling.

D.

These are only possible questions.

What

What does Jerry believe would be an effective means of improving the market competitiveness of firms?.
Developing an efficient hypothesis testing procedure for PCIS, capable of assessing operational cycle time (OCT) and delivery time (DT) for VLSI.

What information does Becky believe that PDA users could benefit from?
Geographic-based information, for use in cases of emergency.

What has been increasingly emphasized by industry?
Process quality and delivery time.

What is increasingly used everyday?
Geographic-based information.

E.

What can cause policy makers to make inaccurate decisions?
Use of the IMD Competitiveness Simulations.

Who can benefit from Geographical-based information, such as in cases of emergency?
PDA users.

Who will lose their market competitiveness?
Firms that perform poorly in terms of quality and delivery.
What will the lack of an effective performance index and an objective procedure lead to?

Inefficiency and a high overhead cost

What do the IMD Competitiveness Simulations neglect?

Interactions among related factors.

## A.

### Situation 1

A face model can be developed to formulate efficiently the 3D image of an individual's from three 2D images. Three 2D images can be obtained simultaneously using three standard cameras. The images can then be transferred to a personal computer. Additionally, the proposed model can be used to formulate digitally the 3D face. As anticipated, the proposed model can reduce the time required to construct a 3D face by 10%, by optimizing the 3D model, rather than manually retrieving 2D images. The proposed model can minimize the tolerable errors associated with constructing a digital face, enhancing multimedia or animation applications by reducing formulation costs and creating more realistic digital objects.

### Situation 2

An appropriate evaluation model to select brands of natural gas buses can be developed. Brands of natural gas buses can be selected by the proposed model, according to a cost effectiveness analysis. Each criterion can also be evaluated in terms of cost and effectiveness. Two methodologies of multiple attribute decision making (MADM), the technique for order preference by similarity to ideal solution (TOPSIS) and the analytic hierarchy process (AHP), can then be used to rank comprehensively all viable alternatives to bus systems. As anticipated, the ranking methodology can provide a more objective outcome, including weights of related decision groups, than can other methodologies. The ranking methodology can also provide a more flexible procedure with respect to an outcome's complexity. Moreover, the ranking outcome enables decision makers to identify the order preferences for alternatives. The proposed model can also evaluate the relationship among cost and effectiveness of all viable alternatives to bus systems. Furthermore, the proposed model can provide a valuable reference for government when selecting brands of bus systems.

### Situation 3

Cesium iodine and cesium bromide crystals doped with OH impurities can be grown using different methods, including the Stockbarger method in vacuum and the Kyropolous method in air. Gamma irradiation can be performed at room temperature, using a Co60 gamma-source at a dose rate of 3800 rad.s-1.The conductivity of both irradiated and non-irradiated crystals can be measured as a function of temperature. As anticipated, the conductivity can be explained by the different thermal stabilities of simple and complicated radiation defects, as well as by accompanying recombination processes. The results of this study can provide further insight into the peculiarities of rebuilding radiation defects in cesium halides doped with OH impurities.

## B.

NOTE: The following are possible questions.

Why does Patrick want to develop a face model?
To formulate efficiently the 3D image of an individual' sface from three 2D images.

Why can Sally's proposed model provide a valuable reference for government?
Because it selects brands of bus systems.

## C.

NOTE:  The following are possible questions.

What does Jim plan to use to perform gamma irradiation at room temperature?
A Co60 gamma-source at a dose rate of 3800 rad.s-1.

What does the ranking outcome of Sally's proposed methodology enable decision makers
to do?
Identify the order preferences for alternatives.

What can Sally's proposed model evaluate?

The relationship between the cost and effectiveness of all viable alternatives to bus systems.


D.

How much time can Patrick's proposed model save in constructing a 3D face?
10%

How does Jim plan to explain conductivity in his research?
By the different thermal stabilities of simple and complicated radiation defects as well as by accompanying recombination processes


E.

What can be grown using different methods?
Cesium iodine and cesium bromide crystals doped with OH impurities

What does Jim believe that results in his study can provide further insight into?
The peculiarities of rebuilding radiation defects in cesium halides doped with OH impurities.

What does Jim want to measure as a function of temperature?
The conductivity of both irradiated and non-irradiated crystals.

# Answer Key
Writing the abstract (part one): briefly introducing the background, objective and methodolody
摘要撰寫（第一部分）：簡介背景、目標及方法

## A.

### Situation 1

Although Taiwan's Ministry of Education is increasingly emphasizing the use of multi-assessment in middle schools and universities, the conventional methods of assessing students' abilities fail to assess higher-order thinking since they can not motivate students properly. Combining the flexibility of a network with the storage capacity of a computer, this work presents a novel networked portfolio system that uses peer assessment to assess the higher-order thinking of students.

### Situation 2

Although an increasing number of Genetic Algorithm (GA) courses on solving optimization problems are offered, students spend much time in coding programs as exercises when learning GAs, many of which can not be implemented in a short time. Therefore, this work presents a novel learning environment that can assist students in flexibly learning genetic algorithms, based on computer-assisted instruction. Several benchmark problems are integrated in this environment. A mathematical expression is then developed to provide fitness functions of GAs. Next, a case study of a GA course is presented to demonstrate the effectiveness of the proposed environment.

### Situation 3

Taiwan's global competitiveness ranking in the IMD World Competitiveness Scoreboard is falling. The IMD Competitiveness Simulations neglect interactions among factors, possibly causing policy makers to make inaccurate decisions. Therefore, this work develops a novel competitiveness model that can integrate a global optimization algorithm into the IMD World Competitiveness Model. Population data are collected from the IMD World Competitiveness Yearbook. Variables are then analyzed.

## B.

NOTE: The following are possible questions.

Does Ann's novel learning environment assist students in flexibly learning genetic algorithms?
Yes.

Does Richard's network portfolio system combine the flexibility of a network with the storage capacity of a computer?
Yes.

Do the IMD Competitiveness Simulations consider interactions among factors?
No.

## C.

NOTE: The following are possible questions.

What may using IMD Competitiveness Simulations cause policy makers to do?
Make inaccurate decisions.

What do students of genetic algorithm courses spend much time in doing?
Coding programs as exercises when learning GAs

What is Matt's methodology to achieve his objective?
Population data are collected from the IMD World Competitiveness Yearbook. Variables
are then analyzed.

## D.

NOTE: The following are possible questions.

Why does Richard want to develop a networked portfolio system that uses peer assessment?

To assess the higher-order thinking of students.

Why does Ann present a case study of a GA course?

To demonstrate the effectiveness of the proposed environment.

Why does Richard want to develop a networked portfolio system that uses peer assessment?Because conventional methods of assessing students' abilities fail to assess higher-order thinking.

E.

What is Ann's novel learning environment capable of?

Assisting students in flexibly learning genetic algorithms.

Why does Ann develop a mathematical expression?

To provide fitness functions of GAs.

What do students spend much time doing when learning GAs?

Coding programs as exercises.

## Answer Key

Writing the abstract (part two): summarizing the anticipated results of the project and its overall contribution to a particular field

摘要撰寫（第二部分）：歸納希望的結果及其對特定領域的貢獻

### A.
### Situation 1

Simulation results indicate that the proposed model reduces the time to construct a 3D face by 10%, by optimizing the 3D model rather than manually retrieving 2D images. Moreover, the proposed model minimizes the tolerable errors associated with constructing a digital face, enhancing multimedia or animation applications by reducing formulation costs and creating more realistic digital objects.

### Situation 2

Simulation results indicate that the proposed architecture can automatically page PDA users through their handheld mobile devices when desired information becomes available. Moreover, the novel architecture allows PDA users to access information according to their geographic position, which functions as a filter. Our results further demonstrate that the architecture proposed herein is highly promising for commercialization as PDA software.

### Situation 3

Analytical results indicate that the novel learning environment reduces the time required for students to complete a GA assignment to one week, increasing the number of exercises to be practiced and improving the learning of GAs. Moreover, the learning environment eliminates the need for hand coding GA programs, simplifying the process of learning genetic algorithms.

## B.

NOTE:  The following are possible questions

How does Melody's novel architecture allow PDA users to access information?
According to their geographic position.

How can Melody's proposed architecture automatically page PDA users?
Through their handheld mobile devices.

How can William's proposed model enhance multimedia or animation applications?
By reducing formulation costs and creating more realistic digital objects.

## C.

NOTE: The following are possible questions.

By what percentage can William's proposed model reduce the time to construct a 3D face?
10%.

What do William's simulation results indicate?
The proposed model reduces the time to construct a 3D face by 10%, by optimizing the 3D model rather than manually retrieving 2D images.

What do Melody's results further demonstrate?
The architecture proposed herein is highly promising for commercialization as PDA software.

## D.

NOTE:  The following are possible questions.

**Unit Six**

*Answer Key*
Writing the abstract (part two): summarizing the anticipated results of the project and its overall contribution to a particular field
摘要撰寫（第二部分）：歸納希望的結果及其對特定領域的貢獻

Is Melody's proposed architecture highly promising for commercialization as PDA software?
Yes.

Does William's proposed model increase the tolerable errors associated with constructing a digital face?
No.

Can students using Sherry's novel learning environment complete a GA assignment in one week?
Yes.

E.

What does Melody's novel architecture allow PDA users to access according to their geographic position?
Information.

Whom can Melody's proposed architecture automatically page?
PDA users.

What does Sherry's learning environment eliminate?
The need for hand coding GA programs.

## About the Author

Born on his father's birthday, Ted Knoy received a Bachelor of Arts in History at Franklin College of Indiana (Franklin, Indiana) and a Masters of Public Administration at American International College (Springfield, Massachusetts). He is currently a Ph.D. student in Education at the University of East Anglia (Norwich, England). Having conducted research and independent study in Ukraine, South Africa, India, Nicaragua, and Switzerland, he has lived in Taiwan since 1989 where he is a permanent resident.

An associate researcher at Union Chemical Laboratories (Industrial Technology Research Institute), Ted is also a technical writing instructor at National Tsing Hua University (Department of Computer Science) and National Chiao Tung University (Institute of Information Management and Department of Communications Engineering). He is also the English editor of several technical and medical journals in Taiwan.

Ted is the author of the Chinese Technical Writers' Series, which includes An English Style Approach for Chinese Technical Writers, English Oral Presentations for Chinese Technical Writers, A Correspondence Manual for Chinese Technical Writers, An Editing Workbook for Chinese Technical Writers, and Advanced Copyediting Practice for Chinese Technical Writers. He is also the author of Writing Effective Study Plans, which is part of the Chinese Professional Writers' Series.

Ted created and coordinates the Chinese On-line Writing Lab (OWL) at http://mx.nthu.edu.tw/~tedknoy

## Acknowledgments

Thanks to the following individuals for contributing to this book:

### National Tsing Hua and National Chiao Tung Universities (Taiwan)

Jyh-Da Wei, Ming-Da Wu, Ing-Hong Liao, Yi-Min Kao, Professor Chuen-Tsai Sun, Professor Han-Lin Li, Professor Lee-Eeng Tong, Professor Shyan-Min Yuan, Professor Wen-Hua Chen, Zhih-Feng Liu, Dr. San-Ru Lin, Chou-I Chin, Min-Juen Dzhen, Min-xin Shen, Professor Duen-Ren Liu, Hung-Hui Chen, Hung-Je Ju, Ming-Sheng Yeh, En-Jie, Li, Chang-Chi Yu, Wan-Gye Yang, Pei-Fang, Yeh, Cheng-Chang Liang,  Ya-Lan Yang, Ming-Wei Chang, Hsih-Tien Chen, Tai-Ing Yeh, Hu-Min Chu, Jung-Fa Tsai, Chang-Jui Fu, Kuo-Tzahn Chang, and Chiu-Hsia Hsieh.

### Institute of Condensed Matter Physics at the Ukrainian Academy of Sciences and Institute of Physical Studies at Ivan Franck University(Ukraine)

Kyrylo Taburshchyk, Oleg Velychko, Maxym Dudka,  Viktoria Blavats'ka, Roman Bigun, Oleh Klochan, Oleg Bovgyra, Ihoz Kajun, Tryna Kudyk, Roman Meruyk, Iryna Yaraphn, Iryna Hud, Oksama Mel'nyk, and Peter Pandiy

### Silicon Integrated Systems Corporation, Hsinchu Science-based Industrial Park (Taiwan)

吳曉韻 鄔志賢 姜兆聲 游曜聲 邱美淑 劉銓　 林平偉 朱俊威
曾若媚 韓承諺 蕭志成 陳方松 林宏洲 蔡璧禧 林柏廷 許資力
邱首凱 邱榮槺 葉國煒 呂忠晏 溫俊福 金愼遠 吳江龍 李鳳笙
林玫君 陳怡安 潘眞眞 李杏櫻 羅惠慈 孫蓮　 吳忠儒 賴東明
林志聰 翁育生 王志浩 陳仲一 林俊宏

Thanks also to Wang Min-Chia for illustrating this book.  Thanks also to my technical writing students in the Department of Computer Science at National Tsing Hua University, Institute of Information Management, Department of Communication Engineering, and Department of Information Science at National Chiao Tung University, Institute of Physical Studies at Ivan Franck University (Lviv, Ukraine) and the Institute of Condensed Matter Physics of the Ukraine Academy of Sciences (Lviv, Ukraine). Robin Koerner and Seamus Harris are also appreciated for reviewing this workbook.

# 精通科技論文（報告）寫作之捷徑
## An English Style Approach for Chinese Technical Writers （修訂版）

作者：柯泰德（Ted Knoy）

## 內容簡介

使用直接而流利的英文會話

讓您所寫的英文科技論文很容易被了解

提供不同形式的句型供您參考利用

比較中英句子結構之異同

利用介系詞片語將二個句子連接在一起

**萬其超 / 李國鼎科技發展基金會秘書長**

本書是多年實務經驗和專注力之結晶，因此是一本坊間少見而極具實用價值的書。

**陳文華 / 國立清華大學工學院院長**

中國人使用英文寫作時，語法上常會犯錯，本書提供了很好的實例示範，對於科技論文寫作有相當參考價值。

**徐　章 / 工業技術研究院量測中心主任**

這是一個讓初學英文寫作的人，能夠先由不犯寫作的錯誤開始再根據書中的步驟逐步學習提升寫作能力的好工具，此書的內容及解說方式使讀者也可以無師自通，藉由自修的方式學習進步，但是更重要的是它雖然是一本好書，當您學會了書中的許多技巧，如果您還想要更進步，那麼基本原則還是要常常練習，才能發揮書的精髓。

**Kathleen Ford, English Editor, Proceedings(Life Science Divison),**
**National Science Council**

The Chinese Technical Writers Series is valuable for anyone involved with creating scientific documentation.

※若有任何英文文件修改問題，請直接與柯泰德先生聯絡：（03）5724895

特　　價　新台幣300元

劃　　撥　19419482 清蔚科技股份有限公司

線上訂購　四方書網 www.4Book.com.tw

發 行 所　華香園出版社

# 作好英文會議簡報
## English Oral Presentations for Chinese Technical Writers

作者：柯泰德（Ted Kony）

## 內容簡介

本書共分十二個單元，涵括產品開發、組織、部門、科技、及產業的介紹、科技背景、公司訪問、研究能力及論文之發表等，每一單元提供不同型態的科技口頭簡報範例，以進行英文口頭簡報的寫作及表達練習，是一本非常實用的著作。

### 李鍾熙／工業技術研究院化學工業研究所所長

一個成功的科技簡報，就是使演講流暢，用簡單直接的方法、清楚表達內容。本書提供一個創新的方法（途徑），給組織每一成員做為借鏡，得以自行準備口頭簡報。利用本書這套有系統的方法加以練習，將必然使您信心倍增，簡報更加順利成功。

### 薛敬和／IUPAC台北國際高分子研討會執行長
### 國立清華大學教授

本書以個案方式介紹各英文會議簡報之執行方式，深入簡出，為邁入實用狀況的最佳參考書籍。

### 沙晉康／清華大學化學研究所所長
### 第十五屆國際雜環化學會議主席

本書介紹英文簡報的格式，值得國人參考。今天在學術或工商界與外國接觸來往均日益增多，我們應加強表達的技巧，尤其是英文的簡報應具有很高的專業水準。本書做為一個很好的範例。

### 張俊彥／國立交通大學電機資訊學院教授兼院長

針對中國學生協助他們寫好英文的國際論文參加國際會議如何以英語演講、內容切中要害特別推薦。

※若有任何英文文件修改問題，請直接與柯泰德先生聯絡：（03）5724895

特　　價　新台幣250元
劃　　撥　19419482 清蔚科技股份有限公司
線上訂購　四方書網 www.4Book.com.tw
發 行 所　工業技術研究院

# 英文信函參考手冊
## A Correspondence Manual for Chinese Technical Writers

作者：柯泰德（Ted Knoy）

## 內容簡介

本書期望成為從事專業管理與科技之中國人，在國際場合上溝通交流時之參考指導書籍。本書所提供的書信範例（附磁碟片），可為您撰述信件時的參考範本。更實際的是，本書如同一「寫作計畫小組」，能因應特定場合（狀況）撰寫出所需要的信函。

**李國鼎／總統府資政**

我國科技人員在國際場合溝通表達之機會急遽增加，希望大家都來重視英文說寫之能力。

**羅明哲／國立中興大學教務長**

一份表達精準且適切的英文信函，在國際間的往來交流上，重要性不亞於研究成果的報告發表。本書介紹各類英文技術信函的特徵及寫作指引，所附範例中肯實用，為優良的學習及參考書籍。

**廖俊臣／國立清華大學理學院院長**

本書提供許多有關工業技術合作、技術轉移、工業資訊、人員訓練及互訪等接洽信函的例句和範例，頗為實用，極具參考價值。

**于樹偉／工業安全衛生技術發展中心主任**

國際間往來日益頻繁，以英文有效地溝通交流，是現今從事科技研究人員所需具備的重要技能。本書在寫作風格、文法結構與取材等方面，提供極佳的寫作參考與指引，所列舉的範例，皆經過作者細心的修訂與潤飾，必能切合讀者的實際需要。

※若有任何英文文件修改問題，請直接與柯泰德先生聯絡：（03）5724895

特　　價　新台幣250元
劃　　撥　19419482 清蔚科技股份有限公司
線上訂購　四方書網 www.4Book.com.tw
發 行 所　工業技術研究院

# 科技英文編修訓練手冊
## An Editing Workbook for Chinese Technical Writers

作者：柯泰德（Ted Knoy）

## 內容簡介

要把科技英文寫的精確並不是件容易的事情。通常在投寄文稿發表前，作者都要前前後後修改草稿，在這樣繁複過程中甚至最後可能請專業的文件編修人士代勞雕琢使全文更為清楚明確。

本書由科技論文的寫作型式、方法型式、內容結構及內容品質著手，並以習題方式使學生透過反覆練習熟能生巧，能確實提昇科技英文之寫作及編修能力。

### 劉炯明 / 國立清華大學校長

「科技英文寫作」是一項非常重要的技巧。本書針對台灣科技研究人員在英文寫作發表這方面的訓練，書中以實用性練習對症下藥，期望科技英文寫作者熟能生巧，實在是一個很有用的教材。

### 彭旭明 / 國立台灣大學副校長

本書為科技英文寫作系列之四；以練習題為主，由反覆練習中提昇寫作反編輯能力。適合理、工、醫、農的學生及研究人員使用，特為推薦。

### 許千樹 / 國立交通大學研究發展處研發長

處於今日高科技時代，國人用到科技英文寫作之機會甚多，如何能以精練的手法寫出一篇好的科技論文，極為重要。本書針對國人寫作之缺點提供了各種清楚的編修範例，實用性高，極具參考價值。

### 陳文村 / 國立清華大學電機資訊學院院長

處在我國日益國際化、資訊化的社會裡，英文書寫是必備的能力，本書提供很多極具參考價值的範例。柯泰德先生在清大任教科技英文寫作多年，深受學生喜愛，本人樂於推薦此書。

※若有任何英文文件修改問題，請直接與柯泰德先生聯絡： (03) 5724895

特　　價　新台幣350元
劃　　撥　19419482 清蔚科技股份有限公司
線上訂購　四方書網 www.4Book.com.tw
發 行 所　清蔚科技股份有限公司

# 科技英文編修訓練手冊【進階篇】
## Advanced Copyediting Practice for Chinese Technical Writers

作者：柯泰德（Ted Knoy）

## 內容簡介
本書延續科技英文寫作系列之四「科技英文編修訓練手冊」之寫作
指導原則，更進一步把重點放在如何讓作者想表達的意思更明顯，
即明白寫作。把文章中曖昧不清全部去除，使閱讀您文章的讀者很
容易的理解您作品的精髓。
本手冊同時國立清華大學資訊工程學系非同步遠距教學科技英文寫
作課程指導範本。

**張俊彥 / 國立交通大學校長暨中研院院士**
> 對於國內理工學生及從事科技研究之人士而言，可說是一本相當有用的
> 書籍，特向讀者推薦。

**蔡仁堅 / 前新竹市長**
> 科技不分國界，隨著進入公元兩千年的資訊時代，使用國際語言撰寫學術
> 報告已是時勢所趨；今欣見柯泰德先生致力於編撰此著作，並彙集了許多
> 實例詳加解說，相信對於科技英文的撰寫有著莫大的裨益，特予以推薦。

**史欽泰 / 工研院院長**
> 本書即以實用範例，針對國人寫作的缺點提供簡單、明白的寫作原則，
> 非常適合科技研發人員使用。

**張智星 / 國立清華大學資訊工程學系副教授、計算中心組長**
> 本書是特別針對系上所開科技英文寫作非同步遠距教學而設計，範圍內
> 容豐富，所列練習也非常實用，學生可以配合課程來使用，在時間上更
> 有彈性的針對自己情況來練習，很有助益。

**劉世東 / 長庚大學醫學院微生物免疫科主任**
> 書中的例子及習題對閱讀者會有很大的助益。這是一本研究生必讀的
> 書，也是一般研究者重要的參考書。

※若有任何英文文件修改問題，請直接與柯泰德先生聯絡： (03) 5724895

特　　　價　新台幣450元
劃　　　撥　19419482 清蔚科技股份有限公司
線上訂購　四方書網 www.4Book.com.tw
發 行 所　清蔚科技股份有限公司

# 有效撰寫英文讀書計畫
## Writing Effective Study Plans

作者：柯泰德（Ted Knoy）

## 內容簡介

本書指導準備出國進修的學生撰寫精簡切要的英文讀書計畫，內容包括：表達學習的領域及興趣、展現所具備之專業領域知識、敘述學歷背景及成就等。本書的每個單元皆提供視覺化的具體情境及相關寫作訓練，讓讀者進行實際的訊息運用練習。此外，書中的編修訓練並可加強「精確寫作」及「明白寫作」的技巧。本書適用於個人自修以及團體授課，能確實引導讀者寫出精簡而有效的英文讀書計畫。

本手冊同時為國立清華大學資訊工程學系非同步遠距教學科技英文寫作課程指導範本。

于樹偉／工業環境與安全衛生技術發展中心主任

　　　《有效撰寫讀書計畫》一書主旨在提供國人精深學習前的準備，包括：讀書計畫及推薦信函的建構、完成。藉由本書中視覺化訊息的互動及練習，國人可以更明確的掌握全篇的意涵，及更完整的表達心中的意念。這也是本書異於坊間同類書籍只著重在片斷記憶，不求理解最大之處。

王玫／工業研究技術院／化學工業研究所企劃與技術推廣組組長

　　　《有效撰寫讀書計畫》主要是針對想要進階學習的讀者，由基本的自我學習經驗描述延伸至未來目標的設定，更進一步強調推薦信函的撰寫，藉由圖片式訊息互動，讓讀者主動聯想及運用寫作知識及技巧，避免一味的記憶零星的範例；如此一來，讀者可以更清楚表明個別的特質及快速掌握重點。

※若有任何英文文件修改問題，請直接與柯泰德先生聯絡：（03）5724895

特　　　價　　新台幣450元
劃　　　撥　　19419482 清蔚科技股份有限公司
線上訂購　　四方書網 www.4Book.com.tw
發　行　所　　清蔚科技股份有限公司

## The Chinese
# Online Writing Lab
【 柯泰德線上英文論文編修訓練服務 】
http://mx.nthu.edu.tw/~tedknoy

您有科技英文寫作上的困擾嗎？
您的文章在投稿時常被國外論文審核人員批評文法很爛嗎？以至於被退稿嗎？
您對論文段落的時式使用上常混淆不清嗎？
您在寫作論文時同一個動詞或名詞常常重複使用嗎？

您的這些煩惱現在均可透過柯泰德網路線上科技英文論文編修
服務來替您加以解決。本服務項目分別含括如下：

    1. 英文論文編輯與修改
    2. 科技英文寫作開課訓練服務
    3. 線上寫作家教
    4. 免費寫作格式建議服務，及網頁問題討論區解答
    5. 線上遠距教學（互動練習）

另外，為能廣為服務中國人士對論文寫作上之缺點，柯泰德亦
同時著作下列參考書籍可供有志人士為寫作上之參考。

    ＜1.精通科技論文（報告）寫作之捷徑
    ＜2.做好英文會議簡報
    ＜3.英文信函參考手冊
    ＜4.科技英文編修訓練手冊
    ＜5.科技英文編修訓練手冊（進階篇）
    ＜6.有效撰寫英文讀書計畫

上部分亦可由柯泰德先生的首頁中下載得到。
如果您對本服務有興趣的話，可參考柯泰德先生的首頁標示。

**柯泰德網路線上科技英文論文編修服務**
地址：新竹市大學路50號8樓之三
TEL:03-5724895
FAX:03-5724938
網址：http://mx.nthu.edu.tw/~tedknoy
E-mail:tedaknoy@ms11.hinet.net

# 有效撰寫英文工作提案

著　　者☞ 柯泰德（Ted Knoy）

出 版 者☞ 揚智文化事業股份有限公司

發 行 人☞ 葉忠賢

責任編輯☞ 賴筱彌

登 記 證☞ 局版北市業字第 1117 號

地　　址☞ 台北市新生南路三段 88 號 5 樓之 6

電　　話☞ （02）23660309　（02）23660313

傳　　真☞ （02）23660310

郵撥帳號☞ 14534976

戶　　名☞ 揚智文化事業股份有限公司

法律顧問☞ 北辰著作權事務所　蕭雄淋律師

印　　刷☞ 鼎易印刷事業股份有限公司

初版二刷☞ 2005 年 4 月

Ｉ Ｓ Ｂ Ｎ ☞ 957-818-382-8（平裝）

定　　價☞ 新台幣 450 元

網　　址☞ http://www.ycrc.com.tw

E－m a i l ☞ tn605541@ms6.tisnet.net.tw

國家圖書館出版品預行編目資料

有效撰寫英文工作提案 ＝ Writing effective
work proposals／柯泰德（Ted Knoy）著. --
初版. -- 臺北市：揚智文化，2002[民 91]
　　面；　公分. --（應用英文寫作系列；2）
ISBN　957-818-382-8（平裝）

　1.商業英語 － 應用文

　493.6　　　　　　　　　　　　91003960